EDISON
the man who made
the future

"A novel interpretation of the Master Mind of the Machine Age,"
by the South American artist Rafael Valdivia

EDISON
the man who made the future

Ronald W. Clark

Macdonald and Jane's
London

First published in Great Britain in 1977 by
Macdonald and Jane's Publishers Ltd, Paulton House, 8 Shepherdess Walk,
London N1 7LW

ISBN: 0 354 04093 6

This book was designed and produced by
George Rainbird Limited, 36 Park Street, London W1Y 4DE

House Editor: Felicity Luard
Designer: Pauline Harrison

Printed and bound by Butler & Tanner Limited, Frome and London

The publishers are indebted to the following for permission to
quote from their publications: The Society of Authors
on behalf of the Bernard Shaw Estate (*The Irrational Knot* by G. B. Shaw);
R. Gelatt and Hawthorne Books, Inc. (*The Fabulous Phonograph*
by R. Gelatt, © Appleton-Century 1966); *Harper's Magazine* ("Edison
in his Laboratory" by M. A. Rosanoff, September 1932, © 1932, renewed 1960,
by *Harper's Magazine*); Macmillan Publishing Co., Inc.
(*J. Pierpont Morgan: An Intimate Portrait* by H. L. Satterlee, © 1939 by
H. L. Satterlee, renewed 1967 by Mabel Satterlee Ingalls).

Contents

*Tenniel's cartoon of coal, steam, and the infant electricity,
published in* Punch, *25 June 1881*

"Addled" Youth

Among Edison's first memories was the recollection of three "prairie schooners" drawn up before his parents' house in Milan, Ohio: the covered wagons in which families were setting out for the 2,000-mile trek to California, across plains teeming with buffalo and with Indian warriors still determined to resist the whites. When he was born, in 1847, Lincoln was about to serve in Congress; when he died, in 1931, Herbert Hoover was announcing his moratorium on war debts and reparations.

The life of Thomas Alva Edison thus spanned the making of modern America. It was, moreover, a life which helped give both the United States and Europe the technological sinews of the contemporary world: electrical industries, the viable telephone network, the phonograph, and the movies. Yet these are only the more famous legacies of an inventive genius who on average lodged a patent every two weeks of his adult life. It is in character that thirteen years before Heinrich Hertz's revelation of radio waves Edison should have experimented with the sparks of his "etheric force," should have discovered the "Edison effect," which was to become the mainspring of the electronics industry, and should have failed to follow up either experiment or discovery because he had more urgent things to do.

Around a man of such fertile mind, legends are bound to grow. Edison, whose genius included that of backing into the headlines, did little to discourage them. Throughout a long life he was prodigal with interviews and reminiscences; anxious, and occasionally overanxious, to tell the world what he was doing, a weakness that at times hoist him on his own claims; and perpetually willing to provide the kind of exciting incident his listeners wanted. He rarely invented history; he often embroidered it.

Among the most persistent legends to grow around the story of America's greatest inventor was that of the poor man's son, denied schooling because of extreme poverty, clawing his way up the ladder of fortune like Matthew Arnold's Shakespeare, "Self-school'd, self-scann'd, self-honour'd, self-secure." Like most legends, it has a grain of truth, though not a very large one. The Edisons emigrated from Amsterdam in the 1730s and finally settled on the Passaic River in New Jersey. Little is known of their early years in America although the head of the family was a banker in Manhattan. Their fortunes, such as they were, seeped away through a combination of good principles and bad luck. John Edison, the inventor's grandfather, took the British side in the War of Independence and had eventually to seek refuge in Nova Scotia. From there he finally moved west to Canada and settled at Bangham on Lake Erie. John's son, Samuel, read what he thought was a wise lesson from his father's experience. John had supported the King and lost his home as a consequence, a choice that his son was not going to make. When the Canadian Rebellion of 1837 broke out, Samuel supported the insurgents. Once again, the family found itself on the losing side and the Edisons were obliged to flee back to the United States.

By the early 1840s Samuel Edison had established a modest but prosperous timber business at Milan, a small town a few miles from the southern shore of Lake Erie. His home was a single-story house with attic rooms, built on a hillside with basement opening out at a lower level, unpretentious and typical of those occupied by respectable small merchants. His wife Nancy, a Canadian woman of Scottish descent, had borne four children before the family arrived in Milan. Three more were to follow, but of the first six children three died in the early 1840s and the survivors were aged fourteen, sixteen, and eighteen when she gave birth to her seventh child on 11 February 1847. He was named Thomas after an ancestor and Alva after Captain Alva Bradley, a Great Lakes shipowner and friend of the family.

Around the boy there later grew the tanglewood of stories and recollections which genius frequently attracts. He was publicly beaten by his father for setting fire to the family barn. He fell into the canal, nearly died in a grain elevator and was remembered—years later, when it was necessary to remember something about the famous Thomas Edison—as the lad who was always in trouble. There was also the occasion when he went swimming with another boy, became separated from him, and returned alone. His companion was later found drowned, and it has been suggested that for the rest of his life Edison suffered from a repressed feeling of unnecessary guilt at the accident—even that his grim dedication to work was a side effect of the incident. It seems unlikely. Edison appears in honest retrospect to have had a youth very much like most other boys of his time. What distinguished him from the rest was an outsize bump of curiosity, an instinct to test the truth of what people told him, and a double dose of energy and impertinence.

The first upheaval came at the age of seven. In 1854, the opening of the Lake Shore Railroad along the southern side of Lake Erie forced the operators of the Milan canal to reduce their tariffs. Business patterns began to change and the Edisons moved to Fort Gratiot on the outskirts of Port Huron, more than 100 miles north of their previous home and on the southern tip of Lake Huron. With this move the story of the poor Edisons takes on a measure of truth. But although Samuel no longer owned his own home he still carried on a business in the lumber and grain trade. No longer the comfortably-off trader of Milan, he was not the implied pauper of some Edison legends.

Soon after the move north Tom Edison caught scarlet fever and it was not until 1855, at the age of eight and a half, that he began attending the white school house. Here he showed what has almost become a sign of genius: after only three months he returned home in tears, reporting that the teacher had described him as "addled." This was in fact no cause for alarm. Leonardo da Vinci, Hans Andersen and Niels Bohr were all singled out in their youth as cases of retarded development; Newton was considered a dunce; the teacher of Sir Humphry Davy commented, "While he was with me I could not discern the faculties by which he was so much distinguished"; and Einstein's headmaster was to warn that the boy "would never make a success of anything." As youths, all had one characteristic in common: each was an individualist, saw no need to explain himself and was thus listed among the odd men out.

Whether Nancy Edison took the Port Huron schoolteacher's opinion seriously or whether she rightly believed herself a better teacher than the local man is a moot point. But Edison remembered the outcome for the rest of his life.

> I found out what a good thing a mother was, she brought me back to the school and angrily told the teacher that he didn't know what he was talking about. She was the most enthusiastic champion a boy ever had, and I determined right then that I would be worthy of her, and show her that her confidence had not been misplaced.

Family loyalty and affection no doubt played its part in recollection. Nevertheless, it is from the age of seven, when mother takes over, that the story of the young Edison begins to diverge from that of his contemporaries. By the age of nine he had read Richard Green Parker's *Natural and Experimental Philosophy* and at the age of thirteen he discovered the writings of Thomas Paine on his father's bookshelves. Almost three-quarters of a century later he wrote:

> It was a revelation to me to read that great thinker's views on political and theological subjects. Paine educated me then about many matters of which I had never before thought. I remember very vividly the flash of enlightenment that shone from

Paine's writings. ... My interest in Paine and his writings was not satisfied by my first reading of his works. I went back to them time and again, just as I have done since my boyhood days.

He also struggled through Newton's *Principia*. It was important in giving him the respect for practice rather than theory which explains both his success as an inventor and the slowness with which the world of science recognized his achievements. Newton's masterpiece, however, helped to give him an almost arrogant contempt for mathematics, an attitude not entirely balanced by his skill in seeing intuitively to the heart of many problems based on figures. A family friend helped to explain the *Principia* in simple language. The result, according to Edison was:

[I] at once came to the conclusion that Newton could have dispensed his knowledge in a much wider field had he known less about figures. It gave me a distaste for mathematics from which I never recovered ... I look upon figures as mathematical tools which are employed to carve out the logical result of reasoning, but I do not consider them necessary to assist one to an intelligent understanding of the result.

Years afterward he was to claim: "I am not a mathematician, but I can get within 10 percent in the higher reaches of the art." And, with more truth but more arrogance: "I can always hire mathematicians, but they can't hire me."

That Edison was to reach a position in the financial pecking order so much above that of most mathematicians was due largely to his practice of relentless practical experiment, begun in the basement of his Port Huron home in the 1850s as he worked his way through the pages of *Natural and Experimental Philosophy*, repeating the author's experiments to confirm that what Parker said would happen, actually did happen.

For a boy of his native energy and instinctive curiosity about the natural world, it was a good decade in which to be young. During the early years of the century André Marie Ampère, Karl Friedrich Gauss and Georg Ohm had begun to disentangle the links connecting electricity and magnetism, first definitively shown by Hans Christian Oersted's earlier observation of the deflection of a compass needle by a nearby electric current. Michael Faraday suggested that the phenomena of electricity and magnetism could best be considered in terms of fields, or areas of space over which their forces were exercised, and James Clerk Maxwell was soon preparing his revolutionary theory that neither electricity nor magnetism existed in isolation, and working at his equations which linked the two phenomena.

Building on this theoretical work, inventors had already harnessed the newly understood force of electricity to create the glaring arc light, limited in use but already giving spectacular results, as in Paris where the Place de la Concorde was lit by the luminous discharge between sputtering carbon rods. In America, where the arc was already blazing out its glaring light, Samuel Morse had patented an electromagnetic device by which messages in the dots and dashes of his eponymous code could be sent along metal wires. In 1844 the words "What hath God wrought" had been tapped out along a wire between Washington and Baltimore and three years later New York had been connected by the telegraph to Washington.

All these achievements had come, if not directly from the scientists who had shown them to be theoretically possible, then at least from their close collaborators. Research and development, in its modern sense, had been minimal or nonexistent, and sheer wonder at the arc light or the telegraph had been enough to push them into immediate use. If this was one explanation of the primitive quality which typified the equipment of the emerging technological world, another was that the development of materials and techniques for specific scientific purposes was still in its infancy. Any young man of the 1850s, glimpsing even a small part of the landscape then opening up, could well have felt like Kipling's explorer who

> heard the mile-wide mutterings of unimagined rivers,
> And beyond the nameless timber saw illimitable plains.

Edison was lucky. As it was written in 1926, had he been born into early twentieth-century America, "he would probably have become professor of a technical institute, or a technical employee in a Trust," impeded by regulations and financial dependence from achieving greatness.

In the less disciplined, less hidebound, world of the 1860s, he had first to find out what was really known; to understand and to observe; to be perpetually asking the most difficult question of all—"why?" He did this, and more, as he repeated Parker's textbook experiments in the cellar of his parents' Port Huron house. He scrounged bottles from the local shopkeepers and later remembered that there were eventually 200 of them, each filled with a different chemical. Near explosions and near disasters almost inevitably followed. Most were looked upon by the boy's parents not only with worry and a fear about what might happen next, but also pride that their boy could even begin to understand such things. "Addled," indeed!

He soon grew ambitious. The Grand Trunk Railroad had in 1859 completed a line from Portland, Maine, to Sarnia, on the eastern side of the St. Clair River, opened a car ferry between Sarnia and Port Huron on the western bank of the river, and simultaneously completed a single-track line which ran south from

Port Huron to Detroit. This suggested possibilities and Edison eventually persuaded his mother to let him apply for a post as newsboy on the morning train from Port Huron to Detroit. He partly persuaded her that he was really grown up, it is asserted, with the announcement: "Ma! I'm a bushel of wheat! I weigh eighty pounds." The story, though told by Edison himself, has an apocryphal ring; nevertheless, like most such stories, it is based on an underlying foundation. It might not be correct as one of Edison's hagiographers was later to claim, that even at this early age his "mind was an electric thunderstorm rushing through the fields of truth." But by the time he was twelve he had begun to quantify the facts of everyday life.

The morning train left Port Huron at 7:00 A.M. and covered the sixty-three miles to Detroit in four hours; left for home at 5:30 P.M. and arrived back at 9:30. To the young Edison the advantages of working the $14\frac{1}{2}$-hour day were twofold. Although as newsboy he would earn only what he could make from paper sales, the concession of candy butcher went with the job, and to a boy of Edison's imagination there was no limit to what he might earn from sales of sweets and food. There was also the six hours' stopover in Detroit which could be spent in the reading room of the Young Men's Association, soon to be reorganized as the Detroit Free Library.

He quickly began to show a sure business instinct. On the morning trip into the city he sold fruit and other local produce from Port Huron, doing it so successfully that he soon had other boys working for him. The evening papers were sold on the trip back, and within a few months profit was running at twenty dollars a week.

Next, he was supplementing business with vegetables from the large garden surrounding his parents' home. "After being on the train for several months," he once said, "I started two stores in Port Huron—one for periodicals, and the other for vegetables, butter and berries in the season. They were attended by two boys who shared in the profits."

He had barely launched this venture when he began to go deaf. According to the earliest published accounts of his life it was the result of a severe boxing round the ears by the train guard, a story which he appears at one time to have endorsed. Later in life he said that he had been delayed by a group of newspaper customers and the train had begun to pull away from the platform without him.

> I ran after it and caught the rear step nearly out of wind and hardly able to lift myself up, for the steps in those days were high [he recalled]. A trainman reached and grabbed me by the ears and as he pulled me up I felt something in my ears crack and right after that I began to get deaf ... If it was that man who injured my hearing, he did it while saving my life.

First there was earache, then a slight difficulty in hearing, finally a deafness that was to become permanent and worse as he grew older. The real cause is still something of a mystery. The aftereffects of scarlet fever have been suggested, so has inheritance, and it is certainly true that Edison suffered from ear troubles throughout a long life, and was twice operated on for mastoids.

It is a measure of his character that he turned to advantage what for most boys would have been a major handicap. Since he was still able to converse with people without too much trouble, since it was the subtleties of music rather than its basic sounds which escaped him, it is clear that the deafness was qualified. Nevertheless, it was hardness of hearing—possibly a more accurate description than deafness in the early days—that drove him to find consolation in the Detroit library even more regularly. Here, he would often recall, he started with the first book on the bottom shelf and worked his way along until the shelf was finished and he could start on the next one.

The disability was also to help with his professional work, even during his early days as a telegraph operator. "While I could hear unerringly the loud ticking of the instrument," he later recalled, "I could not hear other and perhaps distracting sounds. I could not even hear the instrument of the man next to me in a big office."

Later still, when he was developing Bell's early telephone, his difficulty in hearing its faint sounds convinced him of the need for improvements; the outcome was the all-important carbon transmitter, an essential of the instrument even today.

As for the gramophone, Edison was to have no doubts. "Deafness, pure and simple, was responsible for the experimentation which perfected the machine. It took me twenty years to make a perfect record of piano music because it is full of overtones. I now can do it—just because I'm deaf." There was, moreover, one other advantage in defective hearing. In the business jungle where Edison necessarily carried out much of his business he could not rely on verbal agreements; everything had to be in writing, a safety net in what has been called "a business era notorious for financial swindle and brigandage."

At the time of the early deafness he was a plumpish round-faced youngster, a boy of twelve much like his companions except for inexhaustible energy and an audacity that usually overrode opposition. He used a spare freight car on the daily run as his own traveling laboratory and, without a by-your-leave, used the same car as his own printing works after acquiring a hand printing press and begging sufficient type from a friend on the *Detroit Free Press*. The result, appearing soon afterward, was the *Grand Trunk Herald* of which Edison printed 400 copies a week. According to some stories, laboratory and press were eventually thrown off the car after the chemicals had caused a fire.

13

My copy [he would recall] was so purely local that outside the cars and the shops I don't suppose it interested a solitary human being. But I was very proud of my bantling, and looked upon myself as a simon-pure newspaper man. My items used to run like this: "John Robinson, baggage master at James's Creek Station, fell off the platform yesterday and hurt his leg. The boys are sorry for John." Or it might be: "No. 3 Burlington engine has gone into the shed for repairs."

There were also more exciting items as when he reported, under "Births": "At Detroit Junction G.T.R. Refreshment Rooms on the 22nd inst., the wife of A. Little, of a daughter."

Edison's *Grand Trunk Herald* was an omen of many things to come since it brought him into the newspapers for the first time. George Stephenson, the British engineer, made an extensive inspection of the line and was reported by the London *Times* as having commented on the Edison publication—the first newspaper to be produced on a train as he described it.

Shortly afterwards there came an incident that illustrated three traits that were to run through the whole of Edison's life and that were symptomatic of his age. They were quickness at turning chance circumstance to his own benefit, a refusal to be deterred, and a relentlessness—some would say a ruthlessness—in exacting as much payment for a job as the traffic would stand.

One of his main problems had been to estimate accurately how many papers he would sell on the Detroit–Port Huron run each evening. If he carried too few he could lose business; if he carried too many he could end up with unsold stock. To minimize the risk he persuaded a compositor on the *Detroit Free Press* to show him a proof of the paper's main news story each day. Thus warned, he could estimate what the sales on the home run were likely to be.

Then, on an afternoon in early April 1862, his friend showed him the proof of a sensational front page story. The Civil War was already a year old and now Grant and Sherman had met in a huge and bloody battle at Shiloh near Corinth, Tennessee. The Confederates had lost General Albert S. Johnston and although the battle was still raging and the issue still in doubt, killed and wounded were already reported to number 25,000.

"I grasped the situation at once," Edison recalled. "Here was a chance for enormous sales, if only the people along the line could know what had happened. If only they could see the proof slip I was then reading! Suddenly an idea occurred to me." First he made for the telegraph operator on the Detroit station. Would he, Edison asked, telegraph to each of the main stations down the line and suggest that the station master should chalk up the news of the battle on the boards usually carrying the train times. In return Edison offered to supply the man with *Harper's Weekly, Harper's Monthly*, and an evening paper for the next six months. That bargain struck, he went to the *Free Press* offices and asked for 1,500 copies on credit. On being refused he talked his way into the office of the editor, Wilbur F. Storey, who listened to his request in silence. Then he handed the boy a slip of paper, saying: "Take that downstairs and you will get what you want."

I took my fifteen hundred papers, got three boys to help me fold them, and mounted the train, all agog to find out whether the telegraph operator had kept his word [Edison remembered]. At the town where our first stop was made I usually sold two papers. As the train swung into that station, I looked ahead, and thought there must be a riot going on. A big crowd filled the platform, and as the train drew up I began to realize that they wanted my papers. Before we left I had sold a hundred or two at five cents a piece. At the next station the place was fairly black with people. I raised the ante, and sold three hundred papers at *ten* cents each. So it went on until Port Huron was reached. Then I transferred my remaining stock to the wagon which always waited for me there, hired a small boy to sit on the pile of papers in the back of the wagon, so as to discount any pilfering, and sold out every paper I had at a quarter of a dollar or more per copy. I remember I passed a church full of worshippers and stopped to yell out my news. In ten seconds there was not a soul left in the meeting. All of them, including the parson, were clustered round me, bidding against each other for copies of the precious paper.

You can understand why it struck me then that the telegraph must be about the best thing going, for it was the telegraphic notices on the bulletin boards which had done the trick. I determined at once to become a telegraph operator.

Yet, there was more to it than that. The Civil War had broken out just as the telegraph had begun to revolutionize civilian life, and it had quickly become evident to both sides that it could also revolutionize war. As the commanders of the maneuvering armies began to realize how the telegraph both enlarged their sources of intelligence and strengthened their links with officers in the field, first scores, then hundreds, of telegraph operators became attached to the marching and countermarching forces. The Union Army alone eventually had 1,500 on its payroll. As the demand rose the supply diminished. Thus the profession had a doubly romantic attraction: participation in a war in which both sides fought for strongly held beliefs, and in a new science with a great future.

Thus a man of Edison's push and verve might well have become an operator whatever the chances of Fate. As it was, he took the first step following a storybook illustration of the young boy-hero at work.

A few months after the Battle of Shiloh had brought him a relative fortune, the mixed train from Port Huron to Detroit, carrying both freight and passengers, stopped at Mt. Clemens for the half-hour during which shunting had to take place. Edison was already friendly with the red-bearded station agent, J. U. Mackenzie, and Mackenzie's two-and-a-half-year-old son, Jimmy.

The train, of some twelve or fifteen freight cars, had pulled ahead and had backed in upon the freight-house siding, had taken out a box car (containing ten tons of handle material for Jackson State prison), and had pushed it with sufficient momentum to reach the baggage car without a brakeman controlling it [Mackenzie later recalled]. Al, who had been admiring the fowls in the poultry yard, happened to turn at this moment and notice little Jimmy on the main track, throwing pebbles over his head in the sunshine, utterly unconscious of the danger he was in. Al dashed his papers (which was under his arm) upon the platform, together with his glazed cap, and plunged to the rescue, risking his own life to save his little friend, and throwing the child and himself out of the way of the moving car. They both landed face down in sharp, fresh gravel ballast with such force as to drive the particles into the flesh, so that, when rescued, their appearance was somewhat alarming.

Mackenzie, like most other station agents, spent his salary before he got it and was unable to show his gratitude in the usual way. Instead, he offered to teach Edison telegraphy. The boy came daily, cutting short his railway trip. Within a few weeks he was more proficient than Mackenzie himself.

Chapter Two

Tramp
Telegraphist

Edison's first job as a telegraph operator arrived through a combination of good luck and the local reputation he had achieved by the age of sixteen. The day operator of the small telegraph office at Port Huron was anxious to join the Military Telegraph Corps with its prospects of excitement and high pay, but felt that he must get a replacement before leaving. Edison, the protégé of Mackenzie and a bright lad known to make a success of whatever he tackled, was an obvious candidate. He got the job without difficulty and for the next few months became an almost permanent feature of the office, taking and sending messages during the day, then staying on until the early hours of the morning as the press reports came in and were taken down by an experienced operator.

From the employer's viewpoint, he was distinctly casual at the job. His aims were not only to become a proficient telegraphist but to understand how the principles of electricity were being applied. In the nature of things the daytime work was intermittent and when neither sending nor receiving messages he could usually be found in the basement, repeating experiments he had read about in the *Scientific American.* If customers' messages sometimes had to wait a while that was too bad.

There were advantages of the Port Huron post quite apart from the happy-go-lucky way in which Edison was allowed to treat the work. The office formed part of a jeweler's shop and the watchmaker's tools which lay about were ideal for tinkering with homemade telegraph equipment. There was also the freedom with which he was allowed to stay on at the office until the small hours, "cutting in" on the experienced operator who took the press reports. These reports demanded more than the shorter run-of-the-mill messages, and the ability to take "press" without mistakes was the hallmark of the capable man. But before the end of 1863 Edison had learned all they could teach him at Port Huron. Now, and once more with Mackenzie's help, he successfully applied for a post as railway operator with the Grand Trunk Railroad at Stratford Junction, about a hundred miles east across the Canadian frontier. Here he went in the fall, the first move in six years of restless wanderings, which did so much to make him the man he became.

The life of the "tramp operators" in the 1860s was rough and adventurous. They were genuine freelances, selling their skill to the highest bidder, neither having nor wanting security, and often moving from one job to the next on the spur of the moment. Telegraphy itself was hardly out of its swaddling clothes, great financial empires were soon to be built up on its development, and it is not surprising that many of the men who ended the century as captains of industry had learned to survive in the rough-and-tumble of the itinerant operator's life. Nevertheless, there was another side to the coin. If the profession attracted the ambitious, it also attracted the feckless, those content with a world of bedraggled working conditions and seedy rooming houses.

At Stratford, Edison worked the night shift, devoting the day to his own private business of experiment and study. As a result he needed, on duty, all the rest he could organize. This involved circumventing the "six signal," which had to be tapped out every hour as a sign that the operator was awake at his post, a requirement presenting no difficulty to the future maker of the incandescent bulb and the gramophone. He constructed a notched wheel, attached to a clock, which every hour made the necessary connections and sent out the "wide-awake" signal.

He survived discovery of the device. What he did not survive was failure to pass on orders to hold a freight train, which was narrowly saved from a head-on collision. Worried about the consequences, he fled to Sarnia, on the Canadian frontier, and took the ferry to the Port Huron side.

These early troubles were only the first of many during his six years of wandering apprenticeship. "A classic example of what every manufacturer knows: a good research man usually makes a poor operator on the production line," Edison rated opportunities for experiment far higher than routine duties. Combined with youthful lightheartedness and almost a contempt for authority, it brought frequent reprimands and, almost as frequently, dismissal.

Back in Port Huron, looking for a job, Edison gave a telling example of his ability to find simple solutions to problems that baffled most other people. The winter had been severe and the masses of ice clogging the river between Port Huron on one side and the Canadian city of Sarnia on the other eventually severed the cable that linked them. The ice had already made the mile-and-a-half-wide river impassable; with the cable broken, the two "neighboring" cities were effectively cut off from each other.

While the inhabitants were wondering what could be done Edison argued himself into one of the Port Huron locomotives on a line near the riverbank. Then, pulling on the whistle cord, he began sending the dots and dashes of the Morse code across the frozen river in the shrill blasts of the engine whistle. "Hello, Sarnia; Sarnia, do you get what I say?"went the question. At first there was no response. There was none to a second attempt. Then someone in Sarnia realized what was happening and the watchers on the American bank heard the Morse reply from a Canadian engine. Communication between the two cities had been restored.

Early in 1864 Edison found a job with the Lake Shore & Michigan Southern Railroad at Adrian, sixty miles southwest of Detroit. As usual he took the "night trick" which most operators disliked, since it gave him more leisure to experiment. It also gave him time to read—the *Police Gazette* one moment and the *Journal of Higher Mathematics* the next.

At Adrian his downfall came through obeying orders. Told to break in on a line to send an important dispatch, he did so against the protests of the man at the other end. He soon found he had been arguing with the superintendent and was promptly fired for having done what he was told.

Next came a spell in Fort Wayne. But all he could get was a day job and after two months he moved on to Indianapolis where he joined Western Union, the corporation with which his later fortunes were to be tightly bound up.

Edison had by this time evolved his own distinctive sending style, remembered years later by Robert Underwood Johnson, the journalist and author who had worked for a while in 1864 as part-time telegraphist.

> A memorable experience of this episode, which lasted hardly a year, was to listen for what might be called the autograph of a certain operator in the B office at Indianapolis named Edison. The telegraphic style of the great inventor that was to be was unique, and was detected by its lightning-like rapidity. It was the despair even of expert telegraphers, who often had to break into his narrative to ask him to repeat.

Although this was also a day job, Edison and another operator would turn up uninvited at the office after hours and help out the night man by taking press reports. The night man, who liked both his drink and the chance of sleeping it off, was only too happy, and for some weeks the less experienced youths coped with the reports, each receiving messages for ten minutes at a time, writing out as much as he could, then continuing with the rest from memory after the other man had taken over. This worked well enough until, at the other end of the line, Western Union put on an operator with a speed too fast for the youngsters to handle. "He was one of the quickest dispatchers in the business," Edison later remembered, "and we soon found it was hopeless for us to try to keep up with him. Then it was that I worked out my first invention, and necessity was certainly the mother of it."

The solution sounds simple enough today.

> I got two old Morse registers, and arranged them in such a way that by running a strip of paper through them the dots and dashes were recorded on it by the first instrument as fast as they were delivered from the Cincinnati end, and they were transmitted to us through the other instrument at any desired rate of speed or slowness. They would come in on one instrument at the rate of forty words a minute, and we would grind them out of the other at the rate of twenty-five.

Success lasted for some weeks. But with messages coming at forty words a minute and being transliterated at little more than half that speed, a stockpile quickly accumulated. This could be worked off without trouble under normal conditions. But one night a Presidential vote kept the news pouring in without a break, and the two young operators found themselves falling an hour behind, then an hour and a half, finally almost two hours. Complaints began to come in, an investigation was ordered by the company, and the Morse repeater was banned.

Shortly afterward, in February 1865, Edison moved on once more, this time to Cincinnati where he was again employed by Western Union. It was here, where it appeared to a colleague that most of the time he was monkeying with the batteries and circuits, and devising things to make the work of telegraphy less irksome that he graduated from "plug" operator to the status of "first-class man," able to take press copy for as long as required. His money was raised from $80 a month to $125 and one more step was climbed up the ladder.

He soon moved on again, first to Nashville, then to Memphis. The city was still under military occupation following the end of the Civil War, and he found the chief operator in the telegraph office anxious to restore broken lines of communication; in particular to link up New York and New Orleans again. But it was Edison who achieved the feat in the early hours of one morning, doing the trick with what appears to have been an improvised autorepeater in which a message produced by a telegraph receiver can be fed into a transmitter on a different line. The *Memphis Advertiser*, with an office in the same building, discovered what was happening, and published a story on the news. But when Edison arrived at the office the next morning he was fired without explanation. The reason may not have been merely the annoyance of a superior outclassed by a junior. In later life Edison had a cockiness which was the outward sign of a splendid optimism and a belief that difficulties existed merely to be overcome; even then, not everyone liked it and before the rough corners had been rubbed off he must at times have been unbearable.

It was possibly in Memphis that Edison first met the irrepressible George Gouraud who a decade later was to become his agent and propagandist in London. After Edison's death a friend of Gouraud, who had died some years earlier, recalled how at the end of the Civil War the Colonel had called at a telegraph office to find the operator asleep but the machine giving the automatic "sixing" signal for which Edison had already been in trouble elsewhere. The story is garbled, but it seems possible that Gouraud did meet his later employer in some such circumstances. If so, Memphis would be the most likely place.

Edison's next port of call was Louisville, where he stayed for two years before drifting up to Detroit and then back again to Louisville. It was here, he said later, that he developed the unusual vertical style of writing that served him well for the rest of his life. The young Edison found that imagination was needed to fill in the gaps in messages, and to exercise imagination he needed time. The actual writing therefore had to be done at speed and he quickly evolved his small and distinctive upright letters, each separated from the next, and lacking any flourishes. As he was taking an average of eight to fifteen columns of news reports every day it did not take him long to perfect the method.

It was during this second spell in Louisville, Edison once admitted, that he was badly caught out.

> Down in Virginia [he said] the Legislature was trying to elect a United States senator. John M. Botts was the leading candidate. But he never received quite enough votes to elect him. Day after day, the sessions dragged along. One day news came that the opposition to Botts was going to pieces and that he would undoubtedly be elected the next day. The next day, just as a dispatch from Richmond began to come, the wire "broke" just as I had received the name "John M. Botts." I took a chance and wrote out a dispatch to the effect that Botts had been elected. The Louisville papers printed it. The following day, they printed a correction. Botts hadn't been elected. The Legislature, as usual, had only adjourned for the day.

When sacked—and it is possible that the mistake over the would-be Senator Botts brought dismissal—Edison moved on without much apparent rhyme or reason. Yet from each job he picked some new piece of expertise, gradually building up a body of experience which enabled him to cope with the unexpected emergencies which were a constant feature of the young art or craft of telegraphy. On one occasion, his voracious appetite for new experiences nearly led him out of the United States to South America after, with two colleagues, he had read an advertisement from the Brazilian Government asking for operators. The three youths traveled to New Orleans where they found the ship in which they planned to sail had been seized by the government as a troop transport, and were left kicking their heels in the city while waiting for the next departure.

Edison's two colleagues waited for the boat. He changed his mind and eventually made his way back to Louisville where the familiar pattern was repeated. He worked hard on the night shift, sandwiched in the four or five hours' sleep that was all he usually needed, and spent most of the day either reading or experimenting with the equipment on which he spent his spare money. The experiments came first, and now led to his downfall. He tipped over a carboy of sulphuric acid, which ran out and went through to the manager's room below. The next morning he was told that the company wanted operators, not experimenters. He could take his pay and get out.

This time he ended up in Cincinnati again. Once more he worked nights. Once more he rented a small room where he could experiment during the day. And now his ingenuity was given an unusual test. One of the Cincinnati operators was George Ellsworth, a man who had worked during the Civil War for John Morgan, one of the Southern guerrilla leaders. Ellsworth had a considerable technical knowledge of the telegraph system and of what could be done with it. During the fighting he had tapped wires for Morgan, sent false messages over Federal lines, and generally shown how the new art of telegraphy could be used to mislead or confuse an enemy. But if he could tap other people's messages, other people could tap his, and he now turned to Edison, suggesting that if some method of sending untappable messages over the wires could be discovered then they could sell it to the government.

Edison did the trick although how he did it is not clear. But Ellsworth disappeared soon after it had been completed and Edison, characteristically, seems to have lost interest once the technical problem had been solved.

Soon afterwards, he left Cincinnati for home in Port Huron, but before the end of 1868 had moved on east to Boston. He still went as a telegraph operator but with such a break from his previous experiences that it is difficult not to believe he felt one stage of his apprenticeship to have been completed. "Go West, young man," might still be the best advice for one type of pioneer. For Edison, it was the East, with its proliferating cities and its expanding business empires, that beckoned with the possibilities held out by the exploitation of electricity.

He arrived in Boston rather as the penniless Dick Whittington arrived in London or Benjamin Franklin arrived on the Philadelphia waterfront after his four-day journey from New York. He had traveled from Port Huron via Toronto, through a raging blizzard which had marooned the train for twenty-four hours in a snowdrift and made it four days late in Montreal. He had little money to cope with the delay and he finally reached Boston illfed, exhausted and almost penniless. An old operator friend, Milton Adams, had heard of a vacancy in the Western Union's Boston office and had paved the way for him to have it. However, Edison still had to be hired. Luckily, the local superintendent sensed the ability that the rough clothes concealed and asked when the new recruit could start. The recruit answered "Now" and was ordered to report for duty that evening.

At the best of times Edison was quite untroubled about his appearance, and into old age was never happier than when he could discard his coat, take off his tie, roll up his sleeves, and get down to work in the shabbiest clothes available. As a young man, carelessness in dress and lack of interest in the pastimes common to most youths gave him an awkward air, a touch of the country cousin in the bustle and hurry of a city telegraph office. What followed might therefore have been expected.

When he reported for duty he was given a pen and assigned to the wire bringing in news from New York. There he waited for an hour before being ordered to another table and told to take a special report for the *Boston Herald*. What he did not know was that the night operators in Boston had been waiting until the fastest man in New York had become available to send the dispatch.

The New York telegraphist started slowly, then gradually increased his speed. Thereupon Edison reduced the size of the characters in which he was taking down the message. This, he had estimated, enabled him to cope with up to fifty-five words a minute, rather faster than anyone could then send.

Soon the New York operator increased his speed, to which I easily adapted my pace [said Edison]. This put my rival on his mettle, and he put on his best powers, which, however, were soon reached. At this point I happened to look up, and saw the operators all looking over my shoulders, with their faces shining with fun and excitement. I knew then that they were trying to put a job on me, but kept my own counsel and went on placidly with my work, even sharpening a pen at intervals, by way of extra aggravation. The New York man then commenced to slur over his words, running them together, and sticking the signals; but I had been used to this style of telegraphy in taking report, and was not in the least discomfited. Finally, when I thought the fun had gone far enough, and having completed the special, I quietly opened the key and remarked: "Say, young man, change off, and send with your other foot."

The battle won, Edison settled down as he had done elsewhere, working nights, studying during the day, and quite willing, even in Boston, to cock a snook at his superiors. Rebuked one day for writing out more than 1,500 words so finely that they had to be copied before being passed on to the newspaper compositors, he reacted strongly. The next messages were written in giant-size letters so big that a single word filled an entire sheet of paper. He was taken off press copy. It is a tribute to his ability that he was not fired.

Meanwhile his studies were becoming more intense. This was partly due to the environment of Boston, a striking contrast with that of the Middle West

23

cities in which he had previously worked. It was encouraged by purchase of Faraday's *Experimental Researches in Electricity*, which revealed to him more fully than before the theoretical foundations on which the practical development of telegraphy had been built. According to one story, possibly apocryphal but as with so many Edison stories exemplifying the essentials, he brought his Faraday back from work one morning at 4:00 A.M., read it until breakfast, then commented to his colleague: "Adams, I have got so much to do and life is so short. I am going to hustle." Years later he believed much the same with his "To stop is to rust," and, "A harvest must be reaped occasionally, not once in a lifetime."

The telegraph was still the most important application of electricity then developed. Its value had been highlighted by the Civil War and in the aftermath of the struggle huge business organizations were rivaling each other in the struggle for its commercial development. Yet the telegraph was only an augury of things to come. In the busy workrooms of the men who made and repaired telegraph instruments there was already a burning belief that the new system could be developed, adapted, improved and expanded to carry out many tasks other than sending messages across hundreds of miles with a speed that would have been inconceivable only a generation previously.

Boston was a center of this restless worrying-away at the possibilities of electricity. It was the ideal place for a young man of Edison's curiosity and energy and he soon became a familiar figure in the workshops of the city. Among them was the factory of Charles Williams, later to become famous as the first maker of Alexander Graham Bell's telephone, and it was here that Edison built the prototype of an automatic vote recorder. It is ironic that this, the first invention of a man who was to patent more than 1,200 and earn a succession of fortunes in the process, should have been a resounding failure.

In his work as a telegraphist, Edison had taken down column after column of Congressional proceedings. He knew, only too well, the lengthy ritual that had to be gone through every time a vote was taken. Each representative's name had to be called separately, and recorded; the vote of yea or nay had to be added to one of two lists, and then the next representative's name was called. The time wasted was immense; to save it, Edison devised a simple piece of apparatus which could be installed beside the desk of each representative, and which included two buttons or switches, one for a "yes" and one for a "no" vote. In front of the Speaker's desk there was a frame containing two dials and on these there appeared, as each member pressed one of his two buttons, the cumulative totals of "yes" and "no" votes.

To handle the patenting of the device Edison hired a well-known Boston patent lawyer, the Honorable Carroll D. Wright, afterward United States Commissioner of Labor, a man who later described his young client as uncouth in manners, a chewer of tobacco rather than a smoker of it, but full of intelligence and ideas. Papers for the vote recorder were applied for in 1868 and taken out some eight months later.

The instrument was first shown to Massachusetts State officials. Their interest was marginal, and the device was next demonstrated before a Congressional Committee in Washington. Here the verdict was blunt but honest. "Young man," went the message to Edison, "if there is any invention on earth that we don't want down here, it is this. One of the greatest weapons in the hands of a minority to prevent bad legislation is filibustering on votes, and this instrument would prevent it." This was no doubt true. However, what would have inhibited the minority would almost certainly have helped the majority and it is difficult not to believe that what really hamstrung Edison's first invention was the bane of the inventor down the ages: man's instinctive dislike of change, even when change means improvement.

Failure with the vote recorder did little to discourage him. In December 1868 *The Telegrapher* announced that his address was "Care of Charles Williams Jr., Telegraph Instrument Maker, 109 Court Street, Boston," and the following month the same journal stated that: "T. A. Edison has resigned his situation in the Western Union office, Boston, and will devote his time to bringing out inventions." However, experience with the politicians had taught him one thing: to concentrate his energies only on inventions for which there was an obvious need.

It was this, as much as anything else, that attracted him to improvement of the stock ticker, the electrical device which kept businessmen up to date with the minute-by-minute fluctuations of prices on the Stock Exchange. In the financial boom following the end of the Civil War the existing practice of using runners to carry news of the latest prices from the Exchange to individual offices had become increasingly unworkable. It was at first superseded by a system in which subscribers to a stock service were provided with stock indicators. These were electrically linked to the service headquarters, incorporated electro-magnets controlling figures on a dial, and showed, simultaneously in a number of offices, the prices being relayed from the headquarters. E. A. Callahan then substituted type wheels for the dials and so designed the equipment that the figures, instead of being shown on a dial, were printed on a long line of tape—the ticker tape whose name came from the distinctive noise made by the machine in action. The early stock printers were primitive in action, often unreliable, and open to an almost endless succession of improvements. In Boston Edison not only devised his own improved version, which printed the letters of the alphabet as well as figures, but opened an agency which was soon serving between thirty and forty subscribers.

25

He also began work in a field where he later gained both fame and fortune, that of multiplex telegraphy or the method of sending more than one message over a single telegraph wire. Moses G. Farmer, the American electrical engineer who was his predecessor in work on the incandescent lamp, had as far back as 1852 devised a duplex system for sending two messages simultaneously. Later inventors had added refinements and by 1869 Joseph B. Stearns had evolved a system that was successful enough to be in experimental use with Western Union. All the systems involved a wiring arrangement very different from that of the normal telegraph in which messages are transmitted by the passage or nonpassage of an electric current through the system. The best systems still had serious operational problems and it was in the hope of removing them that Edison built his own apparatus and, early in 1869, applied for permission to test it over Western Union wires.

Western Union, unimpressed, turned him down. But in the contemporary atmosphere of intense rivalry, one firm's rejection often meant another's acceptance. Edison therefore applied to the rival Atlantic & Pacific Telegraph Company, aroused their interest, borrowed $800 to finish building his equipment, and early in April arrived at the Telegraph Company's offices in Rochester.

He had already fully briefed the operator at the other end of the line in New York and his hopes were high. Anyone reading the following Saturday's issue of *The Telegrapher* would have thought they had been justified since it reported that Edison's duplex machine had been "tried between New York and Rochester, a distance of over 400 miles by wire, and proved to be a complete success." But as was often to be the case with press reports of Edison equipment that had been born under Edison's influence, fact had been transformed into fiction. The tests had gone on for some days. They had been a complete failure and at the end of a week Edison had returned in disgust to Boston. He had no money and few prospects. There remained only the hope of getting a job in New York. He left his few possessions in Boston, borrowed the fare and arrived penniless at his destination. Here, more than in the years of his upbringing, lay the germ of the poor-boy-makes-good story.

Chapter Three

Professional Inventor

When Edison arrived in New York, he found there were no immediate vacancies with Western Union. No vacancies for him, that is: it is possible that news of Edison the odd man out, the junior always anxious to improve on his employer's ways of doing things, had filtered up the wires from one of the many places where his precociousness had earned him dismissal. Whatever the reason, until he found work shelter was the urgent need. He found it, by one of the luckiest chances in his life, in the battery room of the Wall Street offices of the Laws Gold Reporting Company.

The company had been formed a few years earlier by Dr. S. S. Laws, vice-president and presiding officer of the New York Gold Exchange. In the aftermath of the Civil War trading in gold had risen to unprecedented levels and the current practice of chalking up the fluctuating price on a blackboard in the "Gold Room" had become unworkable. Dr. Laws had first devised his "gold indicator," a mechanical contrivance high on the wall of the Gold Room, which showed the changing price as it was noted by the Gold Board's registrar. As dealings increased he devised a faster, electrical, method of showing the price fluctuations.

However, the up-to-date price still had to be carried by messenger boys to the individual brokers' offices. To eliminate this Laws conceived the idea of a master transmitter, connected by wire to receivers in individual offices and doing for gold what was already done for stocks. He gained a franchise for the scheme, retired from the Gold Exchange, set up his own company, and by 1869 was serving several hundred offices, each of which received their service from the master machine on Wall Street.

Dr. Law's chief engineer, Franklin L. Pope, had already heard of the stock ticker which Edison had devised in Boston. There appears to have been some contact between the two and Edison called on Pope, presumably seeking a job.

There was no vacancy but Pope, sensing the younger man's ability, suggested that he sleep in the cellar of the Laws headquarters until he found a job. In the meantime he could familiarize himself with the Laws equipment.

Edison, having the run of the place once business ceased, soon grasped the details of the system. Shortly afterward, he was in the office during working hours when the transmitter suddenly stopped.

There then took place the famous scene as runners from scores of brokers crowded into the office, each complaining that transmission had stopped and that something must be done about it quickly. Edison went over to the machine and found that a contact spring had broken and jammed two gear wheels. A few seconds later Dr. Laws himself appeared. So did Pope, but in the turmoil both men appear to have lost their heads. Then Laws heard Edison say he knew what the trouble was. He was told to get busy. Edison removed the spring, set the contact wheels at zero, and in about two hours things were working again.

He was then summoned to Dr. Law's office, thoroughly cross-examined about his knowledge of the machines, and told to return the following morning. When he presented himself he was offered the job of helping Pope. The next month Pope resigned and set up on his own as a consulting electrical engineer. Edison took his place, at what was to him the extraordinarily high salary of $300 a month.

His position was now dramatically different from what it had been less than two months earlier when he had arrived penniless in New York. It was to change again, almost as dramatically, during the next three months. In the summer he applied for a patent on an improved version of the stock ticker he had invented in Boston, did the same for yet a third version he worked on with Pope, and adapted the Laws indicator so that its range was as good as that of its rival, the Gold & Stock Telegraph Company.

That company was in the process of being taken over by Western Union, which now found itself faced with an unexpectedly strong competitor. True to its policy, Western Union bought out the competitor and Edison was once again an employee of the company. Shortly afterward he resigned to set up in business with Pope and J. N. Ashley, the publisher of *The Telegrapher* in whose columns it was announced that Pope, Edison & Company were available as Electrical Engineers and as a General Telegraphic Agency with offices in Exchange Buildings, 78 and 80 Broadway. The services offered were extensive. A leading feature would be "the application of electricity to the Arts and Sciences, instruments would be designed to order for special telegraphic services" while special attention would "be paid to the application of Electricity and Magnetism for Fire-Alarms, Thermo-Alarms, Burglar Alarms, etc." In addition the firm offered to draw up telegraphic patents for clients, to supply them with raw materials or finished apparatus, and added: "Our facilities for this business are unexcelled."

Edison was now in a situation ideally tailored to his special abilities. His years as an operator enabled him to understand, almost intuitively, the practical, on-the-ground problems that could upset the use of new equipment that utilized electricity, either for telegraphy or in the closely allied field of the stock ticker. His self-gathered theoretical knowledge gave him at least an idea of what was possible. Finally his experience first in Boston, then with Laws in New York, provided the essential spur: the knowledge of what was wanted.

Within a short while Edison designed his Universal Stock Printer, perfected a new printer, the "gold printer," and organized a service which rented it to subscribers at twenty-five dollars a week. This was a challenge to Western Union's services and, not unnaturally, the company lost no time in buying out the service. As a result Edison earned his first $5,000, his one-third share of the price.

Two developments now followed almost simultaneously. With $5,000 in his pocket, Edison felt free to branch out entirely on his own; and General Marshall Lefferts, a high executive in Western Union, successfully tried to conscript the young man's services exclusively for the company. He began to finance Edison's research, gave him specific jobs to tackle and in what seems to have been a very loose arrangement, saw that the young inventor had little time to work for anyone else.

The situation might have continued for some while had Edison not been given the task of dealing with the habit of stock printers to run wild at times. Occasionally a machine in a broker's office would go berserk, printing figures which had no connection with those being sent from the central office. The practice was for the central office to be informed, after which a mechanic would be sent to inspect and repair the errant machine. Edison's idea, successfully carried out, was that the central office should be able to bring the machine back to normal merely by sending special impulses over the existing wires.

This achievement appears to have convinced General Lefferts that the relationship with Edison should be put on a proper footing. The attention paid to patents by whole sectors of the business community appears to have been concentrated on their evasion; might was often right, and Edison himself was prone to adopt the principle of acting first and asking afterward. However, there were limits to such a state of near anarchy and Edison was now called Lefferts's office.

The interview and its aftermath were later described by Edison. But they are, it must be admitted, a good example of what Matthew Josephson in his monumental biography has called the inventor's tendency "to draw the long bow in his stories, picturing himself as an innocent lamb among the cunning wolves of Wall Street—a lamb who nonetheless made off with more than one bag of gold or greenbacks."

How much, General Lefferts asked, did the young man want for the improvements he had been making to the company's equipment? Edison's story is that he thought they were worth $5,000, but could get along with $3,000, that he lacked the nerve to demand such a huge sum, and asked Lefferts to make him an offer.

The latter may well have been true, but as he had only recently received a one-third share of the $15,000 paid by Western Union for his ticker service, the reason for it was probably tactics rather than lack of nerve.

"How would $40,000 strike you?" Lefferts asked.

"This caused me to come as near fainting as I ever got," Edison recalled. "I was afraid he would hear my heart beat. I managed to say that I thought it was fair."

The contract was duly signed and Edison was given a check for $40,000. He later implied that he had always been paid cash, and that when he tried to pay it into the bank, ignorance of banking methods, plus deafness, prevented him from understanding that it had to be endorsed. The check was refused and for a nasty moment or two he believed he had been swindled. A return to Lefferts resolved the matter and Edison received his money. But the bank clerk, unable to resist perplexing a financial greenhorn, paid the $40,000 in bundles of small bills. Only when Edison explained the situation to Lefferts the following morning was he initiated into the mysteries of opening a bank account. That at least was his story.

Throughout his long life Edison repeatedly stressed that his main interest in making money was to acquire funds for further research and for putting his inventions on the market. That was certainly the case with his first $40,000. Within a few months he had moved into Nos. 10 and 12 Ward Street, Newark, New Jersey, equipped them as workshops and was busy turning out stock tickers and other equipment. Lefferts led off with an order for 1,200 machines, other companies followed, and before many months were out Edison was employing a double shift, supervising both himself, and coping with the problem of sleep by snatching the odd hour or two as and when he could. Soon he had two more premises in the city.

With the establishment of the Newark works Edison became a figure in his own right in the rapidly expanding, wildly competitive and none too scrupulous business world of the Eastern seaboard. As he put it in a letter to his parents, he was now what they thought of as "a bloated Eastern manufacturer." He was, in fact, utterly unlike most of the species. It is true that he soon lost whatever ignorance of banks and banking he had retained; it is true that he had as sharp an eye for a bargain as most men and that his early experiences of roughing it as an operator had both sharpened his wits and made him a shrewd judge of human nature. Yet in one way he was a striking contrast to most of his contemporaries who built up huge fortunes: he had a lordly indifference to such elementary matters as bookkeeping for which his substitute was merely two hooks in the works. "I kept only payroll accounts, no others," he once said; "received the bills, and generally gave notes in payment. The first intimation that a note was due was the protest, after which I had to hustle around and raise the money." The system lasted beyond his Newark days. When his interests expanded and he became the organizer of huge industries he gave in and allowed a business manager to run his affairs more efficiently; but he surrendered only reluctantly.

The Newark factory was a tough place to work in and the team which Edison gathered was forced to adopt his own demanding attitude; the job came first. Many members—the English engineer Charles Batchelor, John Kreusi the Swiss watchmaker, Sigmund Bergman the German mechanic, and John Ott who could turn his hand to most things—were to work with him for years, the nucleus of a group who could tolerate his eccentricities and on whose abilities he depended.

> Mr. Edison had his desk in one corner and after completing an invention he would jump up and do a kind of Zulu war dance [one of his workers said years later]. He would swear something awful. We would crowd round him and he would show us the new invention and explain it to the pattern maker and tell us what to do about it.

All these workers believed. They believed in Edison, and so strongly that whatever disasters threatened they were confident he would pull them back from the brink. He usually did. But despite the extraordinary successes of Edison's early years, they had at times the breath-clutching quality of a Pearl White serial with the outcome in doubt until the last moment.

Less than twelve months after the exciting story of the Newark works had begun, Edison married. It was as well he did so then, since during the next few years he might not have found time. His bride was Mary Stilwell, aged only sixteen. As with so many personal details of Edison's life there are contradictory, and equally authenticated stories of how they first met. However, in the summer of 1871 Miss Stilwell was working in the Newark factory and on Christmas Day, having refused to wait the year that her parents had requested, Edison married his fair-haired bride. He brought her to the new house he had bought in Newark and, shortly afterwards, they left for a honeymoon at Niagara. Legend claims, almost inevitably, that on his wedding day he had to be called away from the laboratory he had installed on the first floor of the factory. A daughter, named Marion, was born the following year; four years later, a son christened Thomas and in 1879 a second son, William Leslie. Edison, still concentrating on the telegraph, nicknamed the first two children, Dot and Dash.

There is no doubt that he was sincerely in love with his wife; indeed, all the evidence points to infatuation. Yet it is difficult not to feel that this was equaled by another feeling; whatever the attractions of home life and the raising of a family, marriage did take the bachelor mechanics of living off his own shoulders and allow him to concentrate, with the minimum of distraction, on the really important business of inventing.

One of the first devices to which he now turned his attention at Newark was the automatic telegraph. The first scheme for automatically sending messages faster than they could be tapped out by hand had first been worked out more than two decades previously by Alexander Bain, and Edison's work in the field was, as so often, the technological improvement of an idea and the removal of the handicaps under which it was then being operated.

The greatest speed at which messages could be sent on the telegraph for any length of time was still only about fifty words a minute. However, Bain had devised a system for hand punching a strip of paper tape with long and short perforations comparable to the dots and dashes of the Morse code. The tape was then automatically drawn between a metal roller and a metal stylus, a circuit being completed whenever the stylus was allowed to make contact with the roller by a long or a short perforation of the paper. Using this device it was possible for tapes to be punched up at leisure and then transmitted at a speed of about 400 words a minute.

During the 1850s and early 1860s various improvements were made to Bain's system but when its most successful development was brought to the United States in the early 1870s automatic telegraphing still had one major disadvantage. Over long circuits—and they were being steadily increased in length— it was found that the individual signals tended to run into each other.

Edison was brought into the story after the best contemporary automatic system, designed by George D. Little, had been bought by a group of businessmen who subsequently set up the Automatic Telegraph Company to operate it. The Little system worked, but it worked badly and in the spring of 1871 one of the machines was brought to Edison in Ward Street by Edward H. Johnson. Would Edison, he asked on behalf of the company, see what could be done about the system they had unwisely bought? Edison, anxious to get his hands on the automatic in any case, grasped the opportunity with alacrity and for $40,000 agreed to assign to the company any improvements that he was able to make.

Early Days of an Inventor

Edison as a newsboy at the age of fourteen

*Edison's parents: Samuel Edison, and
Nancy Elliott Edison*

*ft: Edison's birthplace
 Milan, Ohio. The house, which is
 w a museum, is built on a
 llside and this photograph,
 ken from the rear, shows it
 a two-story building.
 om the street it appears as a
 e-story cottage*

*An artist's fanciful
impression of the
teenage Edison printing
the* Grand Trunk Herald,
*the paper he produced
while working as newsboy
on the Port Huron
Detroit line*

Left: *A typical sheet of
Edison's* Grand Trunk
Herald

Port Huron February 3rd 1862,

ND TRUNK RAILROAD

HANGE OF TIME

Going west.

. leaves Port Huron.7.05 PH
r'Detroit, leaves Pt Huron at 7,40 A.M

GOING EAST.
aves Detroit, For Toronto, at 6 15 A.M
er Pt. Huron, leaves at 4,00 P.M
Freight Train both way.
C. E, Christie, Supt.

STAGES.

EW BALTIMORE STATION
kly stage leaves above named Stat
day for New Baltimore, Algonac, Swan
d Newport,
Graves propietor,

MAIL EXPRESS.
xpress leaves New Baltimore Station
ning on arrival of the Train from Det-
Baltimore, Algonac, Swan Creek, and
Clark & Bennett, prop.

Pt, HURON STATION,
mbus leaves the station for Pt Huron,
ival of all Train
Fare cents, Oley Agent

LOST LOST LOST,
parcel of Cloth was lost on the cars
der will be liberally rewarded.

MARKETS.

New Baltimore Feb 3rd,
Butter at 10 to 12 cts per lb
Eggs, at 12 cts. per doz
Lard at 7 to 9 cents per lb.
Dressed Hogs, at 3.00 to 3.25 per 100 lbs,
Flour—at 450 to 475 per bbl.
Buckwheat at 1.50 per 100 lbs,
Mutton—at 4 to 5 cts per lb,
Beans—at 1.00 to 1.50 per bush.
Potatoes at 40 cts each
Corn at 30 to 35 cts per bush.
Turkeys—at 50 to 65 cts each
Chickens at 10 to 12 cts. a lb.
Geese at 25 to 35 cents each
Ducks at 30 cents per pair.

ADVERTISEMENTS.

RAILROAD EXCHANGE,
At Balitmore Station
The above named Hotel is now open for the reception of Travelers. The Bar will be supplied
with the best of Liquors, and every attention will
be made to the comfort of the Guests.
S. Davis Proprietor

SPLENDID PORTABLE COPYING
PRESSES FOR SALE AT
Mt. CLEMENS ORDERS TAKEN,
BY THE NEWS AGENT ON THE MIXED.

Ridgeway Refreshment Rooms—I would inform
my friends that I have opened a refreshment
room for the accomodation of the traveling public
R, Allen, proprietor,   tf

TO THE RAILROAD MEN
Railroad Men send in your orders for Butter,
Eggs, Lard, Cheese, Turkeys, Chickens, and
Geese   W, C, Hulets, New Baltimore Station,

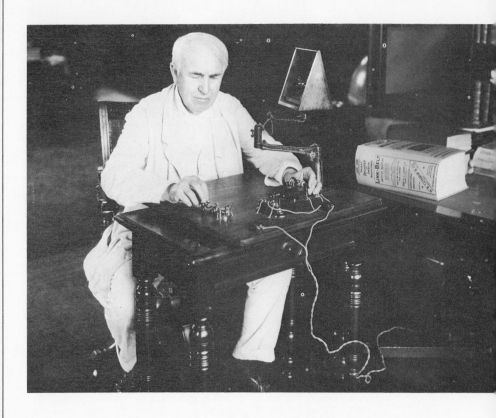

One of the very few
photographs—taken more than
half a century after the Civil
War—showing Edison operating
a telegraph key

Right: *General Grant's
field telegraph station at
Wilcox Landing, near
City Point during the Civil War.
Over this improvised line
Grant received daily reports
from four armies totaling
a quarter of a million men and
gave orders for their
operation over some
750,000 square miles*

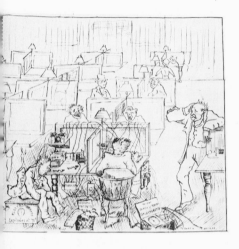

*An artist's impression in 1875 of
the chaos that was—incorrectly—
expected after Edison's development
of the quadruplex telegraph,
which allowed four messages
to be sent over a
single wire*

Below: *Edison's first invention,
a mechanical vote recorder.
After its demonstration before
a Congressional Committee, he was told:
"If there is any invention on earth
that we don't want down here,
it is this"*

The Gold Room in New
York City during "Black
Friday," 24 September 186[
when a group of
financiers tried to corner
the country's gold stocks.
The ensuing panic
emphasized the need for
instruments such as Ediso[
improved gold printer
to record changing prices
with minimum delay

Edison's printing telegraph or Stock Ticker,
made at the Newark works
in the early 1870s

The home of Edison's earliest successful inventions, the Newark factory on Ward Street, 1873

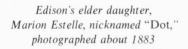

*Edison in the early years
of his inventive career*

*Edison's elder daughter,
Marion Estelle, nicknamed "Dot,"
photographed about 1883*

*Edison's two elder sons,
Thomas A. Edison, Jr. (right), nicknamed
"Dash," and William L. Edison,
photographed about 1883*

Edison's first wife, Mary Stilwell,
photographed in 1871, the year of her marriage
at the age of sixteen

Right: *Alexander Graham Bell in 1876, the year in which the telephone was patented*

Right: *Sir William Preece, the British Post Office engineer—both critic and admirer of Edison's work—being interviewed in his office*

Below: *The Western Electric Company's Multiple Switchboard at the Manchester, England, Telephone Exchange in the 1880s*

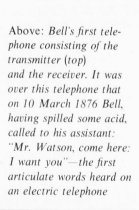

Above: *Bell's first telephone consisting of the transmitter (top) and the receiver. It was over this telephone that on 10 March 1876 Bell, having spilled some acid, called to his assistant: "Mr. Watson, come here: I want you"—the first articulate words heard on an electric telephone*

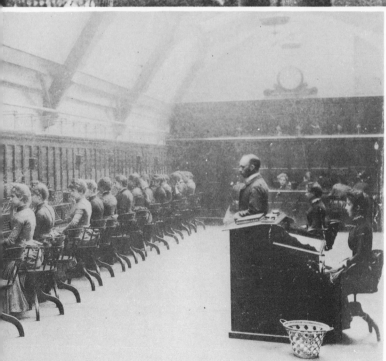

Above: *Edison's electric pen and press, later developed into the mimeograph*
Below: *The century-old original tinfoil phonograph patented by Edison in 1877*

The typewriter, patented
by Christopher Latham Scholes
of Milwaukee, Wisconsin, in 1868 and
developed by Edison at the
request of the Automatic Telegraph
Company during 1871

Left above: The Laboratory at Menlo Park, 1880, Charles Batchelor standing on the steps next to Edison

Left below: The interior of the Menlo Park Laboratory. Edison's father has his back towards the photographer while Charles Batchelor looks out of the window

Edison (right), Charles Batchelor (center), and the well-known journalist Uriah Painter (left), with Edison's first phonograph. A photograph taken during the trip to Washington in April 1878 when Edison demonstrated the phonograph to President Hayes and members of Congress

Preparing to observe the
eclipse at Rawlins, Wyoming, on 29 July 1878:
(from left to right) George F. Baker; Robert M. Galbraith;
Henry Morton; unknown; unknown; D. H. Talbot;
M. F. Rae; Marshall Fox; James E. Watson; Mrs. A. H. Watson;
Mrs. Henry Draper; Henry Draper; Edison;
J. Norman Lockyer

He settled down to work in a typically methodical way and it is significant that he improved the Little machine not by a single stroke but by examining the components of the system, analyzing each of them, and then improving their efficiency, a process revealed by the numerous patents which he lodged during more than two years' work on the undertaking.

One of the first major problems was to produce a chemically prepared paper on which the characters could be recorded at speed. He tackled it in what was to become the typical Edison manner, described by Johnson.

> I came in one night and there sat Edison with a pile of chemistries and chemical books that were five feet high when they stood on the floor and laid one upon the other. He had ordered them from New York, London and Paris. He studied them night and day. He ate at his desk and slept in his chair. In six weeks he had gone through the books, written a volume of abstracts, made 2,000 experiments on the formulas and had produced a solution—the only one in the world—that would do the very thing he wanted done, record over 200 words a minute on a wire 250 miles long.

He overcame the major difficulty, produced by the slow tailing off of each signal into the next, by introducing into the receiver a shunt circuit which, as the current was broken at the end of each signal, created a countercurrent of opposite polarity. This killed the tailing-off effect and resulted in a clean break between each signal. But the actual transmitting tape also had to be improved. Edison abandoned the idea of cutting short and long slits in it to simulate the dots and dashes of the Morse code. Instead, he punched a single circular hole in the tape to represent a dot while two holes with a third hole above and between them represented a dash. In the transmitting instrument the tape was passed between a drum and two parallel wheels. One of the wheels completed the circuit to produce a dot whenever a hole appeared beneath it. When the three holes representing a dash appeared, however, the first wheel made a contact that was continued by the second wheel as the upper dot appeared beneath it, and the contact was continued as the first wheel then made contact through the third of the three holes representing the dash. The result was a clear succession of short dots and long dashes. The effect of the various improvements was cumulative, and just what they meant was demonstrated when a message of about 12,000 words, already punched up, was eventually sent over the line from Washington to New York. Transmission time was $22\frac{1}{2}$ minutes, about 500 words a minute.

By the spring of 1873 the Automatic Telegraph Company had hopes of selling the system to the British Post Office. Edison was obviously the man to clinch the deal and he sailed for England with an assistant, a small satchel of clothes and three large cases of equipment. The visit was not a happy one.

The test set by the Post Office was the transmission from Liverpool to London of 500-word messages, to be sent every half an hour for six hours. Edison's assistant was dispatched to Liverpool; he himself was to handle the receiving set in London. He soon discovered that the equipment he was expected to use was poor, and remedied the defect by buying a new hundred-guinea battery. This was taken to Liverpool just in time for the test which appeared to have passed off satisfactorily. The "time lost" column of the report, Edison later remarked, contained not a single entry.

It was then suggested that even greater speeds might be obtained if his system was used with submarine cables. Eager for any challenge, he was anxious to experiment. For the test he was offered the use of 2,200 miles of cable stored under water in tanks at Greenwich. He had to use it at night, but that was no handicap to a man used to working as much at night as during the day. Describing the experiment, he said:

> I got my apparatus down and set up, and then to get a preliminary idea of what the distortion of the signal would be, I sent a single dot, which should have been recorded upon my automatic paper by a mark about one thirty-second of an inch long. Instead of that it was twenty-seven feet long! If I ever had any conceit, it vanished from my boots up.

Edison worked away for a fortnight, getting slightly better results but nothing that was satisfactory. Here it was his lack of theoretical knowledge which let him down. Owing to induction the cable, coiled in hundreds of loops lying close to each other in the water, inevitably produced poor results. Edison had no idea of this—nor of the fact that his system was giving results in the coiled cable at least as good as those of the existing system.

He returned home disappointed and rather disgruntled, which possibly accounts for his first and slightly jaundiced reactions to Britain. "The English," he said on being asked what his first impressions were, "are not an inventive people; they don't eat enough pie. To invent, your system must be all out of order, and there is nothing that will do that like the good old-fashioned American pie."

In Newark once more, Edison took up again his work on multiplex telegraphy. It had occupied a good deal of his thoughts since his early days as a telegraphist and shortly before the visit to England he had begun working with Western Union on the development of the multiplex patents which they owned.

One of his first successes was the "diplex" telegraph as distinct from the duplex. With the latter, messages could be sent simultaneously from each end of a telegraph wire, but the diplex system allowed two messages to be sent simultaneously from the same end. This was achieved not only by varying the strength of the current to send one message—the method adopted in the duplex system—but by changing the polarity of the current and using such changes to operate a second receiving instrument at the far end of the line.

To a man of Edison's ambition, this was a beginning rather than an end. If two messages could be sent simultaneously in one direction along a single wire, why could not two others be sent at the same time in the other direction? Edison answered the question with the quadruplex.

> This problem was of a most difficult and complicated kind, and I bent all my energies towards its solution. It required a peculiar effort of the mind, such as the imagining of eight different things moving simultaneously on a mental plane, without anything to demonstrate their efficiency.

One of his practical difficulties was to balance the current perfectly, since he was not long in discovering that conventional rheostats were useless for his purpose. Would it be possible, Edison wondered, to make use of the fact that the electrical resistance of some semiconductors varied with the pressure to which they were submitted? To find the answer he mounted a hollow cylinder of insulating material on to a brass plate and placed inside it about fifty silk disks which had been saturated with size, filled with finely ground graphite, and then dried. Above the disks was a second plate upon which a variable and measurable amount of pressure could be applied by means of a screw. The device was wired into the circuit and it was found that when minimum pressure was applied resistance was some 6,000 ohms; if the top plate was screwed down hard on the disks, resistance went down to as little as 400 ohms. The solution, quite satisfactory, was to have great repercussions a few years later in his work on the telephone.

Once having found a way of balancing the currents, Edison continued with his research. It was to go on for some time and during the checkered course of the work he found himself drawn into the first engagements of the war between Western Union and Jay Gould, the most notorious of "the robber barons" fighting for control of America's telegraphs and railroads. He himself gave his official biographers a partial and disingenuous account of the affair. Alvin F. Harlow, in his careful *Old Wires and New Waves: The History of the Telegraph, Telephone and Wireless*, noted a quarter of a century later that "the story of the fight between A & P and the Western Union over Edison's quadruplex sheds an interesting light on Mr. Edison's harum-scarum notions of contractual obligations in those days."

Early in 1873 it appears to have been accepted that Edison would offer any resulting patents to Western Union and that the company would buy them for a sum to be mutually agreed. Following his return from England later in the year the position had changed. His story was that Western Union had lost interest; Western Union's, that Edison felt he lacked the necessary facilities. Both may well be true. What is undeniable is that Western Union's chief engineer, George B. Prescott, was brought in and that under an agreement signed by Edison and Prescott, the latter would be known as joint inventor and would receive some of the profits from any quadruplex device that Edison patented.

However, William Orton, the company's President, maintained there was insufficient security for the payment on account of some $10,000 that Edison urgently needed. And Edison, forced into a financial corner, obtained the money from the Automatic Telegraph Company, apparently on the instructions of Jay Gould, Western Union's greatest rival. Whether Edison knew Gould at the time is not clear; certainly he always tended to play down his connections with one of the most unsavory men of the times. Yet Mary Nerney, Secretary of Historical Research at the Edison Laboratory in the 1920s, has stated that Gould was the owner of one of Edison's rented premises in Newark, and once threatened to sue him for $500. This, she continued, was apparently a move on Gould's part "to get control of Edison's automatic telegraph which he coveted."

Whatever Edison's relations with Gould, he now had $10,000 to pay off outstanding debts. He went on with the quadruplex and by the late autumn of 1874 the time had arrived for a test, to be mounted before the Western Union board, over a wire running from New York to Albany and back to New York. The weather was bad and Edison knew that it could cause trouble. In the circumstances, as he admits, he took as few chances as possible.

"I had picked the best operators in New York, and they were familiar with the apparatus," he later said. "I arranged that if a storm occurred, and the bad side got shaky, they should do the best they could and draw freely on their imaginations. They were sending old messages."

By midday everything began to go wrong, a storm near Albany causing particular trouble. Then William Orton arrived with the other directors.

Despite the storm, the quadruplex passed muster and was soon afterward installed on the lines linking New York with Boston and Philadelphia. Toward the end of December 1874 it appeared that the story was coming to a conclusion. Western Union paid Edison $5,000 on account and it was arranged that they would buy the relevant patents for $25,000 plus a royalty of $233 a year for each circuit on which they were used.

But now came a hitch. According to Edison, Orton left New York and Western Union's general superintendent, General Eckert, warned him that no more money was likely to come from the company. However, Eckert continued, he himself knew of someone who might be willing to buy the quadruplex. This turned out to be Jay Gould, the man whose connection with the

Automatic Telegraph Company had helped to save Edison the previous summer, and it was perhaps natural that Eckert should fail to tell Edison that he was secretly negotiating with Gould to leave Western Union for Gould's own Atlantic & Pacific.

It has been said of Gould that he "thrived on fear, manipulation and panic," and that he "ruined his own friends without scruple and accumulated hatred in a business era notorious for financial swindle and brigandage." Edison's views were ambivalent. "His conscience seemed to be atrophied," he once said, "but that may be due to the fact that he was contending with men who never had any to be atrophied." However, this mild rebuke was counterbalanced by his statement that he never had any grudge against Gould, "because he was so able in his line, and as long as my part was successful the money with me was a secondary consideration."

On 28 December Edison demonstrated the quadruplex to Gould at the Newark works and on 4 January was taken to Gould's home in central Manhattan. Here, no doubt remembering his encounter with General Lefferts a few years earlier, he told Gould: "Make me an offer." This time the price was $30,000 and when he was given a check for the amount the following morning he was told by Gould: "You have got the *Plymouth Rock*," a steamboat he had just sold for $30,000.

A fortnight later Orton, apparently still unaware of what had been going on, sent Edison a letter that was intended to close the quadruplex deal. After two and a half weeks there came a reply.

In it [says Harlow in his summary of the lengthy court case which followed] Edison said that he had made the arrangement with Orton under an error; he found that the quadruplex patents really belonged to one, George Harrington, with whom he had agreed away back in 1871 to share certain automatic patents. An assignment to Harrington was recorded at the time in the Patent Office. Gould had now bought Harrington's interest, and Western Union attorneys found that somebody had forged the word "or" on the Patent Office records to indicate that Edison's agreement with Harrington covered his other telegraph patents.

Western Union lost no time in starting an action against Gould's Atlantic & Pacific which, barred for the moment from employing Edison's quadruplex patents, used one which they claimed had been invented by one of their own employees. It has been claimed to have been an infringement of Edison's. The case, which dragged on for more than a year, did Edison's reputation little good. He was examined in the witness box for eight days and William Preece, the British Post Office engineer, visiting the United States during the case, recorded in his diary:

Orton told me in England of him, "that young man has a vacuum where his con-science ought to be" [a conclusion curiously similar to Edison's on Jay Gould] "and he is known here as [the] Professor Duplicity. The evidence that I heard impli-cated officials in a way that, thank Heaven, is impossible in England. I would not for £50,000 have my name bespattered as Prescott's was. The patent is taken out in the joint name of Prescott and Edison." Edison was asked if Prescott invented any part of the apparatus. "None, sir." "Did he invent anything?" "Never, sir." "Then what has Mr. Prescott to do with it?" "I never could have got it tried if I had not associated with Prescott as joint inventor," and so on and so on.

Edison was of course legally entitled to sell his patents to whomsoever he wished, even to Jay Gould. A freelance offering his services to all comers, he believed—like some nuclear scientists today—that his task was to deliver the goods, whatever the morality of the receiver; that was someone else's affair. He was not yet thirty, he was moving among men whose ethical principles seemed to be nonexistent, and he may genuinely have believed that Western Union would never pay up. Thus Harlow's use of the word "harum-scarum," in its dictionary definition of "reckless, wild, flighty, rash, thoughtless" seems more accurate than "duplicity." There was, moreover, the ethical climate within which Edison moved. "Everybody steals in commerce and industry," he once said to a young industrial recruit. "I've stolen a lot myself. But I knew *how* to steal. They don't know *how* to steal—that's all that's the matter with them." Neverthe-less, Edison paid a price for cutting his corners in negotiations with Western Union. A few years later, when he could have moved, far earlier than he did, against those who were infringing his electric light patents, he shrank from doing so. The eight days in the witness box, explaining how he had worked under Western Union's wing and sold to their opposition, had left their mark.

Some of the $30,000 from Gould no doubt went on tiding Edison over a bad patch. However, in years to come he said:

I spent the whole of it experimenting in trying to make a wire carry six messages instead of four. I didn't succeed. So that financially I am worse off than I would have been had I never invented the quadruplex system.

As for Gould, who was soon to win control of Western Union, his side of the bargain can be judged by Edison's comments on the quadruplex made in the *Scientific American* in 1892:

Inasmuch as every mile of wire actually built does the work of four miles of wire, the quadruplex system represents 216,000 miles of phantom wire, worth $10,800,000.

On these $10 million worth of wires there is no repairing to be done. The value of those phantom wires is, therefore, represented by a saving of $860,000 in repairs at $4 a mile annually, besides the interest on the $10,800,000 which it would have taken to build them.

Before the benefits of Edison's quadruplex were being fully exploited he had carried out another demonstration of his ability to observe constantly, to pigeonhole information, and to pull it from the pigeonhole on demand. In the summer of 1874 the U.S. Federal Court had finally settled the ownership of a patent covering "the use of a retractile spring to withdraw the armature lever from the magnet of a telegraph." This was the Page patent, an essential to the relaying of messages from one circuit to another and necessary on long-distance telegraph lines. Edison has recounted how the loser in the Federal Court came to him in desperation and explained that some way must be found of relaying messages without the use of an electromagnet—an idea considered at the time as comparable to performing *Hamlet* without the Prince. Surely it was only by the use of a magnet that a lever could be moved at the end of a telegraph wire? Only thus, surely, could a current, activating or not activating the magnet, cause a vibrating arm to make or break a circuit and produce the dots and dashes of the Morse message?

It can be claimed that Edison was lucky in readily finding a method of evading the Page patent. It would be fairer to say that success sprang from his habit of never being satisfied until he had reached the "why" of events. In this case he recalled an earlier experiment in which he had noted that if a paper were soaked in certain chemical solutions and a current then passed through them, the resulting electrochemical decomposition made the paper more slippery. The same thing took place, he soon discovered, if a piece of chalk was substituted for the paper.

Starting with this simple fact, he now produced an elegant device. Its essential was a drum of moist chalk, revolved continuously by clockwork. Resting on the chalk was a small pad on the end of a vibrating arm. As the drum revolved the friction between drum and pad carried the pad to the arm's farthest extension: if the friction were eliminated, a spring would draw the arm back. In operation, one pole of a telegraph circuit was connected to the drum, the other to the arm and the pad. When no impulse was being sent over the wire the arm would be kept at full extension; when an impulse, short or long to produce dot or dash, came over the wire, then the arm would fall back. Thus it became a simple matter for the movement of the arm to operate a make-and-break circuit which repeated the incoming impulses—and without the use of an electromagnet.

Quite apart from being a way round the Page patent, the electromotograph, as Edison christened his new device, had one great advantage over the normal magnet-operated apparatus: it would respond to currents so weak that they would fail to operate the conventional receiver. In fact, Edison wrote of one model,

> With a delicately constructed machine, moved by clockwork, which I have nearly finished, I have succeeded in obtaining a movement of the lever, sufficient to close the local circuit with a current (through one million ohms, equal to 100,000 miles of telegraph wire), which was insufficient ... to move an ordinary galvanometer needle.

So far, it had been almost exclusively in this world of telegraphy that Edison had made his mark. But that world was in 1875 on the brink of a revolution. It had been a marvel, difficult for some to believe and even more difficult for many to understand, when Morse had shown that it was possible for messages in the dots and dashes of his code to be sent, instantaneously, over long distances. Even so, it was still far removed from sending the human voice itself. Yet now, in February 1876, Alexander Graham Bell, a Scotsman who had settled in Boston, applied for a patent covering a device which would perform the miracle.

Bell's father had been a pioneer teacher of the deaf, and between 1868 and 1870 the son had helped the father in his Edinburgh school. The family emigrated to Canada and in 1873 Bell moved south to Boston and was appointed professor of vocal physiology at the University. He fell in love with one of his deaf pupils and began to study the mechanical production of sound, being encouraged in this by Joseph Henry, the doyen of American physicists. Bell believed that if the vibrations of sound waves could be transformed into a fluctuating electric current, it might be possible to turn the fluctuations back into sound waves. This was the task done by the device for which he was granted a patent in March 1876. Two months later he read a paper on his invention to the American Academy of Arts and Sciences in Boston, a paper that alerted the telegraph companies to what they immediately recognized as a threat.

Bell's telephone, although a huge stride forward, was a very different instrument from the sophisticated apparatus of today. For one thing, transmission and reception was by means of a single piece of mechanism, which meant that when a message had been spoken into it, the instrument had to be smartly moved up to the ear in order to hear the reply. But there were other limitations inherent in the original telephone which consisted, essentially, of a diaphragm of soft iron enclosed in a mouthpiece and mounted close to the pole of a bar magnet, one end of which was surrounded by a coil of wire. A message spoken into the mouthpiece produced sound vibrations in the diaphragm, which acted as a vibrating armature. This induced electric impulses in the magnetic coil which corresponded to the sound waves; the impulses were carried along a wire to a second

instrument similar to the first; here the impulses produced variations in the magnetism which in turn caused the diaphragm to vibrate and reproduce the original
sounds. But the human voice could generate only weak impulses, and the sounds
produced were thus faint. They were made fainter by the resistance in the wire,
so that Bell's telephone was only effective over comparatively short distances.

Nevertheless, William Orton was well aware of the telephone's implications
for the telegraph companies, and, ignoring the continuing dispute over the multiplex patents, he hired Edison with instructions to put it on a commerical basis.
He did so, and while there is no doubt that Bell invented the telephone, its spread
across the world would have come far more slowly without the basic and revolutionary improvement that Edison now devised.

One of his first accounts of the work was given only two or three years later
to J. B. McClure. Edison had been investigating the possibility of a telephone
for some while before Bell lodged his patent and had early in 1876 sent the Patent
Office a caveat which said that he was working on the subject. However, it was
not until January 1877 that he began to experiment with the varying resistance
of carbon under differing pressures, the phenomenon he had made use of when
balancing currents for his quadruplex system.

> I constructed a transmitter in which a button of some semiconducting substance
> was placed between two platinum disks, in a kind of cup or small containing vessel.
> Electrical connection between the button and disks was maintained by the slight
> pressure of a piece of rubber tubing, half an inch in diameter and half an inch
> long, which was secured to the diaphragm, and also made to rest against the outside
> disk. The vibrations of the diaphragm were thus able to produce the requisite
> pressure on the platinum disk, and thereby vary the resistance of the button in-
> cluded in primary circuit of the induction coil.

Finally he made a small button from the lampblack taken from a smoking
petroleum lamp.

> A small disk made of this substance, when placed in the telephone, gave splendid
> results, the articulation being distinct, and the volume of sound several times
> greater than with telephones worked on the magneto principle.

In Bell's telephone it was the human voice which actually generated the,
necessarily weak, electric current; in Edison's, the human voice was used to open
or close a valve which regulated an electric current of any desired strength. But
there was another feature of the Edison transmitter quite as important in practical terms. In the Bell system it was the original weak currents themselves which
were transmitted along the wire linking one telephone instrument with another.
In Edison's, the battery current was passed through the primary circuit of an

induction coil, the secondary coil of which was able to produce corresponding impulses of enormously higher potential for sending along the wire to the receiver. Thus the range of the telegraph was increased in a single bound from a few miles to hundreds.

These advantages were evident when Edison's transmitter was tested for the first time over a 106-mile line between New York and Philadelphia. It crossed five streams in cables, and in New York City was taken underground before entering the Western Union office with a mass of other wires. In the then state of the art, few tests could have been more rigorous. Nevertheless, the trials were completely successful and every word spoken at one end of the line was clearly heard at the other. In Philadelphia, moreover, a number of differently made receivers were plugged into the line and reception on all of them was found to be equally good.

Edison filed his basic patent for the carbon button telephone transmitter in April 1877 and it is symptomatic of the controversial situation that it was not finally granted until May 1892. Despite the delay in legal recognition, Western Union snapped up the carbon transmitter as soon as its success was obvious. Edison believed it was worth $25,000, but, as with Lefferts and Gould, asked Orton to make him an offer. Orton replied with $100,000. Most men of thirty would have seized the offer without qualification. Edison was different. He accepted, but only on one condition—that he was paid not a lump sum of $100,000 but, instead, $6,000 a year for the next seventeen years, the life of the patent.

Two points are revealing, the first being the reason for Edison's proviso. "My ambition was about four times too large for my business capacity," he later confessed, "and I knew that I would soon spend this money experimenting if I got it all at once, so I fixed it that I couldn't. I saved seventeen years of worry by this stroke." However, he lost $100,000 in interest.

With Edison's carbon transmitter available, Orton could safely start to fight Bell. Soon he set up the American Speaking Telegraph Company with a capital of $300,000 and the power of Western Union behind it. "With all the bulk of its great wealth and prestige," relates Herbert Casson in his history of the telephone, Western Union, "swept down upon Bell and his little bodyguard. It trampled upon Bell's patent with as little concern as an elephant can have when he tramples upon an ant's nest. To the complete bewilderment of Bell, it coolly announced that it had 'the only original telephone' and that it was ready to supply 'superior telephones with all the latest improvements made by the original inventors, Dolbear, Gray and Edison.'" "Original" was of course tendentious, but the statement as a whole appears no worse than many made during the bitter private wars which brought the telephone into general use.

Once the carbon transmitter began to be used, the Bell lessees started clamoring for something as good.

> How to compete with the Western Union, which had this superior transmitter, a host of agents, a network of wires, forty millions of capital, and a first claim upon all newspapers, hotels, railroads, and rights of way—that was the immediate problem that confronted the new General Manager [says Casson]. Every inch of progress had to be fought for. Several of his captains deserted, and he was compelled to take control of their unprofitable exhanges. There was scarcely a mail that did not bring him some bulletin of discouragement or defeat.

While the battle was continuing in the United States, the situation was much the same in Europe, although in each country interpretation of the patent situation could vary. It was in Britain that the war between the Edison and the Bell interests was finally ended. But first there was to be a series of engagements which, serious as they were at the time to those involved, have today an almost comic air as one firm jumps ahead of the other, only to be leapfrogged by its rival a few months later.

It was in the fall of 1878—after Edison had left Newark for new headquarters—that the carbon transmitter was brought to Britain and in November it was tested, with considerable success, between the Norwich factory of J. & J. Colman, and the firm's London office. The test was followed by a public demonstration at the Royal Institution and it seemed that another victory for Edison was in the making.

However, while the Edison transmitter was used at one end of the line in these tests, Bell's magneto-receiver was used at the other. A few months earlier the Telephone Company had been registered in London to exploit the Bell receiver and its British representative, a Colonel Reynolds, pounced immediately: unless Edison stopped using the Bell instrument, action would be taken for infringement. Edison's man in London, Colonel Gouraud, cabled the bad news to America.

Edison, quite unworried, told Colonel Gouraud to delay negotiations. He would devise a new receiver that would circumvent the Bell instrument. Deep as he now was, late in 1878, in efforts to produce an incandescent bulb, he nevertheless acted promptly and ruthlessly. He withdrew all his men from work on the electric light and put them on telephony.

Within three months he had found the answer. It was given by an adaptation of the electromotograph which a few years earlier had provided a way round the Page patent. In the new version the chalk cylinder, revolved either by hand or by a motor, was pressed upon by a spring attached to the receiving diaphragm. The varying electrical current emerging from the transmitter controlled the amount of friction between the chalk and the spring, and this in turn produced vibrations of the diaphragm which repeated the sounds made into the transmitter. Once again, Edison had scored. The new device not only obviated any use of the Bell patent; with it, reception was louder.

Too loud, in fact. George Bernard Shaw, who was to be employed by the Edison company in London as Wayleave Manager, described the device as being "of such stentorian efficiency that it bellowed your most private communications all over the house insted of whispering them with some sort of discretion."

In March 1879 the new receiver, known by a variety of names including the "loud-speaking telephone," the "chalk receiver" and the "motograph receiver," was brought to Britain by Edison's nephew, Charles. The following month it was demonstrated, together with the carbon transmitter, at the rooms of the Royal Society before a distinguished audience that included the Society's President, William Spottiswoode. The transmitter was in the Royal Institution's laboratory in Albemarle Street and a passage from one of Mr. Gladstone's speeches was read over it and clearly received in the Royal Society's rooms in Burlington House some hundreds of yards away.

The receiver, it was announced, had merely been "thrown together in five days" at the urgent request of Colonel Gouraud so that it could be used in a lecture by Professor Tyndall on modern acoustics. And it was, it was added, purely experimental, a phrase possibly intended to lull the Bell camp into believing that nothing serious was intended.

There were, in fact, certain disadvantages about the Edison receiver, one being the need to keep turning the chalk cylinder. Nevertheless, the "electrochemical telephone", using Edison's transmitter and his receiver, was a great improvement on its predecessors and in showing it to the American Association for the Advancement of Science Edison scored a considerable personal triumph.

The demonstration took place in the town hall of Saratoga Springs on 30 August and both Edison and Bell were on the platform. Charles Batchelor, in a nearby room connected by wire to the hall, supplied what the *New York Tribune* called vociferous remarks and thunderous songs. But when these were relayed by Bell's apparatus only one person on the platform could hear the performance, and then only by holding the receiver to his ear. When Edison's carbon transmitter and Bell's receiver was used, a few listeners close to the instrument could hear Batchelor.

> Finally [the paper said] the electrochemical telephone was used with brilliant results. Mr. Batchelor's talk, recitations and singing could be heard all over the hall, and the audience was delighted with such enchanting novelties as "Mary had a little Lamb," "Jack and Jill went up the Hill," "John Brown's Body," "There Was a little Girl" and the like.

Edison, describing the principles of his instrument, was in strong contrast to the sophisticated Graham Bell.

His quaint and homely manner, his unpolished but clear language, his odd but pithy expressions charmed and attracted them [the *Tribune* said]. Mr. Edison is certainly not graceful or elegant. He shuffled about the platform in an ungainly way, and his stooping, swinging figure was lacking in dignity. But his eyes were wonderfully expressive, his face frank and cordial, and his frequent smile hearty and irresistible. If his sentences were not rounded, they went to the point, and the assembly dispersed with great satisfaction at having seen and heard his most recent invention.

A few days before this successful *tour de force* Colonel Gouraud in England had formed the Edison Telephone Company of Great Britain and Edward Johnson had been dispatched to the capital to set up a telephone exchange in opposition to Bell's. In London the telephone was considered as a mixed blessing and *The Times* voiced an opinion that was more widespread than is readily appreciated in the 1970s.

It is a common complaint that the conditions of modern life and especially mercantile life, have been rendered wellnigh intolerable by the telegraph and the addition of the telephone must inevitably "more embroil the fray" [it stated in a leading article]. In old times a man of business could arrange his affairs for the day after the delivery of the morning post, and the perpetual arrival of telegrams has served to add new stings to existence. The case would surely be much worse with the verbal communications than with the written ones.

There would be compensations, notably a reduction in travel and in the traffic congestion of London's streets, but there were strongly conflicting views about the telephone, and Colonel Gouraud had an uphill task.

However, he and his staff were persistent letter writers. Within a few months they informed *Times* readers of a set-piece installation made some fifteen miles outside London.

A gentleman, with whose house the Edison Telephone Exchange has been placed in connection was enjoying a day's hunting [wrote the Edison manager]. At the time he left home the telephone had not been fixed in position, although the wire had been run. During his absence the work was completed, and on his return important letters were read to him to which replies were dictated. A conversation with his solicitors being necessary, the operator at the central exchange connected him in the manner already described in your columns. An important telegram from New York was received, and a reply dictated in time to reach New York three hours after the first message was dispatched from America. The principal portion of a day's work in town was thus compressed into half an hour's accommodation in a library.

By this time the two rival companies, Edison and Bell, were competing vigorously in London. "Their gangs of workmen," says the biographer of William Preece, by now the Chief Engineer of the British Post Office, which was taking great interest in telephone developments, "ran wires over [the City of London's] roofs, put faults on each other's lines, and scrimmaged high above the streets when running near each other." Neither company had any great technical lead over the other since while Bell's company possessed a good receiver and an inferior transmitter, the advantages of Edison's transmitter was partly counterbalanced by the sometimes erratic performance of the receiver.

Much therefore rested on efficient maintenance, and Edison's method of testing his engineers in America before they were sent to England was drastic if successful. First he set up an exchange in which ten of the new telephones were installed and here, before each test, he would himself tamper with each installation, cutting the wires of one, short-circuiting another, putting a third out of adjustment and dirtying the electrodes of a fourth. "When a man could find the trouble ten consecutive times, using five minutes each, he was sent to London," he explained.

Just what the successful technicians were like has been described by George Bernard Shaw:

These deluded and romantic men gave a glimpse of the skilled proletariat of the United States. They sang obsolete sentimental songs with genuine emotion; and their language was frightful even to an Irishman. They worked with a ferocious energy which was out of all proportion to the actual result achieved. Indomitably resolved to assert their Republican manhood by taking no orders from a tall-hatted Englishman whose still politeness covered his conviction that they were relatively to himself inferior and common persons, they insisted on being slave-driven with genuine American oaths by a genuine free and equal American foreman. They utterly despised the artfully slow British workman, who did as little for his wages as he possibly could; never hurried himself; and had a deep reverence for one whose pocket could be tapped by respectful behaviour. Need I add that they were contemptuously wondered at by this same British workman as a parcel of outlandish adult boys who sweated themselves for their employer's benefit instead of looking after their own interest. They adored Mr. Edison as the greatest man of all time in every possible department of science, art and philosophy, and execrated Mr. Graham Bell, the inventor of the rival phone, as his Satanic adversary; but each of them had (or intended to have) on the brink of completion an improvement on the telephone, usually a new transmitter. They were free-souled creatures, excellent company, sensitive, cheerful, and profane; liars, braggarts, and hustlers, with an air of making slow old England hum, which never left them even when, as often happened, they were wrestling with difficulties of their own making, or struggling in no-thoroughfares, from which they had to be retrieved like stray sheep by Englishmen without imagination enough to go wrong.

If the war above the roof tops was raged with some ruthlessness, no less can be said of the propaganda war that went on at the same time. The Bell system, the Edison company stated to its potential customers, "includes a magnet and a coil, and the sound is transmitted along the wire, losing much of its force on the way. In Mr. Edison's instrument ... the speaker is heard with a volume of tone and a distinctness equal to the original utterance." But not, of course, according to Bell, whose company claimed: "The Edison Electrochemical Telephone can scarcely be considered a practical instrument. Its use has been altogether abandoned in the United States and on the Continent, and the authorities in this country find that they cannot make it work satisfactorily"—a statement which one historian of the telephone has sagely noted was "not altogether in accordance with the facts."

In the fall of 1879, the Bell company jumped ahead by acquiring rights in an improved carbon transmitter. It could not be used over long distances, but at a time when the proliferation of the telephone was mainly in city centers this was relatively unimportant and Bell might now have swept ahead had it not been for pressure from the Post Office.

In 1868 and 1869 the British Government had taken over the telegraph companies and transformed operation of the systems into an official monopoly. Bell's telephone had not yet come into existence and such primitive predecessors as existed were, rightly, considered as mere toys. In September 1879, however, the Postmaster-General, Lord John Manners, astonished the telephone companies by announcing that the telephone was a telegraph within the meaning of the Acts of 1868 and 1869, and that private companies would only be allowed to operate under license. Edison and Bell both announced that they would not apply for licenses but it was against Edison that the government moved in 1880. The outcome was a crucial case, "The Attorney-General v. The Edison Telephone Company of London." Edison retained, as expert witnesses, a glittering galaxy of scientists that included Lord Rayleigh, Sir William Thomson, later Lord Kelvin, and Professor John Tyndall. Nevertheless, Edison lost.

It was now obvious that the shadow of Lord Manners loomed over both companies and in the crisis their old argument dissolved. On 8 June 1880 the two companies amalgamated to form the United Telephone Company which took out a thirty-year license from the Post Office in return for 10 percent of its profits. Edison, cabled by Gouraud that he was being offered "30,000" for his interest, accepted at once. When the draft came he was astonished to find it was for 30,000 pounds sterling. He had expected 30,000 dollars.

The saga of the Edison–Bell rivalry which had started with Edison's invention of the carbon transmitter in 1877 thus ended some three years after it had started. During these years Edison had found time to develop numerous ideas, quite apart from his major work on the incandescent bulb which overlapped his controversy with Bell. Among them was the electric pen, the predecessor of the mimeograph or duplicator and an illustration of his ability to produce an invention to fill a specific requirement. As a businessman, even though an unconventional example of the species, he knew how useful it would be if facsimiles of handwritten documents could be produced mechanically—not the one or two copies which could be obtained by the use of carbon paper but copies by the score. For Edison the challenge was met with his usual idea on the back of an envelope, a rough prototype, and then development of a device that could be made in numbers and sold economically. The result was three patents lodged early in 1877 for perforating pens, pneumatic stencil pens and stencil pens.

The initial device was simple enough. A pointed stylus was used to write on a specially prepared tough paper which during the writing was held on a finely grooved steel plate. The process perforated the paper with hundreds of minute holes in an outline of the written message and the resulting stencil was then laid on top of a sheet of ordinary paper. On being inked with a roller it allowed the ink to pass through the holes, thus producing a facsimile copy. The simple stylus was soon superseded by a pneumatically operated needle which moved up and down, punching minute holes as it was passed across the paper, and finally by the end product, the electric pen. This was a slim tube holding a needle-tipped stylus; on top there was mounted a miniature electric motor, whose driven cams operated the stylus many hundred times a minute. The motor was driven by a Bunsen battery of two glass jars incorporating a plunger which would push the plates into the acid solution when the pen was in use and, to save current, withdraw them when it was not. And to make the pen a convenient "desk model" a clip on top of the battery held the pen when it was not in action.

By modern standards, Edison's electric pen appears as the last piece of equipment with which anyone would wish to equip an office. In his day, it was an outstanding success. Up to 3,000 copies could be made from the stencils and within a few years the pens were not only being used in government offices in Washington but exported to countries as distant as Russia and China.

There was also the typewriter. A prototype had been made and patented in 1868 by Christopher Latham Sholes, usually credited as the actual inventor, and two colleagues. It had even more teething troubles than most innovations and more than twenty different models were sent, as improvements were built in, to men likely to be interested in its development. Edison was one of those asked if he could turn it into a commercial proposition. He found it a difficult task. "One letter would be one-sixteenth of an inch above the others," he remembered, "and all the letters wanted to wander out of line. I worked on it till the machine gave fair results." This was what later became the Remington and while Edison, following the familiar pattern, lost interest once the prototype had been made to work, stray thoughts about the machine remained at the back of his mind. "I might invent an electric typewriter, a noiseless one," he commented two decades later. "But the thing is not pressing as it is in very good condition now."

Preoccupation with development of a specific invention to meet a specific need was a feature of Edison's entire working life, and with it there went a contempt, barely concealed at times, for the man who dealt in theories rather than their practical application. Edison well knew that the inventor's world rested on foundations built by scientists. He employed them when necessary, just as he employed mathematicians and metal workers and glassblowers, and he tended to put them on much the same level, rarely acknowledging publicly that the man who worked with his brains should be given a status different from the man who worked with his hands. This was in part the genuine result of his own personal experience which showed that academic training was not necessary for the accomplishment of great things. It was partly the result of a commonsense estimate that in the United States of the 1870s and 1880s there existed a huge reservoir of scientific knowledge that few men had yet thought of turning into technological ironmongery for practical use. There was perhaps a third factor. Edison did genuinely like rolling up his sleeves and working with his hands; almost to the end of his life he could wire a circuit, or for that matter knock in a nail, better than most men. But quite early in his career it had become the picture expected and at times one seems to be watching Edison playing Edison. A delight in practicalities and skepticism of mere theory was an essential of the part.

Nevertheless, there was a real weakness on the theoretical side, and if failure to deal with it had little effect on Edison's ability to bring into the world a record number of devices, methods and processes which made life more worth living, a price had to be paid. Part of it was failure to recognize or exploit the radio waves that he produced twelve years before Hertz's famous experiments, or to explain the "Edison effect" which John Ambrose Fleming was later to utilize in the vacuum tube, the key to the electronics industry.

The first of his two missed opportunities came in November 1875, only a short while before he quit Newark. While experimenting with a vibrator magnet Edison saw a spark coming from the cores of the magnet. "This we have noticed often in relays, in stock printers, when there was a little iron filings between the armature and core, and more often in our new electric pen, and we have always come to the conclusion that it was caused by strong induction," he recorded in his laboratory notebook on 22 November 1875. Now, however, the spark was so strong that he felt another explanation was called for. He then connected the end of the vibrator to a gas pipe and found that he could draw sparks from pipes in any part of the room. Other experiments followed and Edison ended the day's notes with the sentence: "This is simply wonderful, and a good proof that the cause of the spark is a *true unknown force.*"

During the next few weeks he christened the phenomenon "etheric force." It refused to obey most of the established laws of electricity, while it differed from electricity in being independent of polarity, and required neither a circuit nor insulation. It had no apparent effect on the human body unless an exceedingly strong galvanic current was used around the magnet, and none when passed through a prepared frog's leg, the most delicate electric test then known.

While Edison had little idea of what he was producing he recognized the sparks as a signal of energy being transmitted through space, and in the *Operator* of January 1876 elaborated on the possible significance of the phenomenon. "The cumbersome appliances of transmitting ordinary electricity, such as telegraph pole, insulating knobs, cable sheathings, and so on, may be left out of the problem of quick and cheap telegraphic transmission; and a great saving of time and labor accomplished."

Yet his instinctive feel that something could be made of the etheric force was overcome by other emotions. "When I first noticed it," he said years afterward, "I regarded it as a great nuisance. But I was working on so many things at that time that I had no time to do anything more about it. I just observed the results and gave them to others. They went ahead and developed them."

Ten years later, he lodged his only patent covering telegraphy without wires, but it invoked the principle of induction which, he had taken care to show, had nothing to do with the mysterious etheric force.

Chapter Four

The Talking Machine

While Edison was perfecting the electric pen he was also preparing to leave Newark. In later years, he gave more than one reason for the move. To Dyer and Martin he claimed that trouble about rent was the cause. "I had rented a small shop in Newark, on the top floor of a padlock factory, by the month," he said. He moved out after a short time and then found that a local law made him liable for a year's rent. "This seemed so unjust," he went on, "that I determined to get out of a place that permitted such injustice." Some such minor disagreement may have been remembered for a third of a century, but it is significant that much earlier, speaking from Menlo Park, the hamlet where he set up his new head-quarters, he commented: "When the public tracks me out here I shall simply have to take to the woods." And Francis Jehl, his colleague for many years, believed that what he really needed was a more secluded place.

The Newark factory, unconventional though it was, and remarkable as the inventions launched from it undoubtedly were, was a production unit rather than a research laboratory. Edison had by 1876 decided that he wanted something different, an organization which would study contemporary demands and then find methods of meeting them. He had, moreover, made enough money by manufacturing to be sure that he could make a living by invention alone.

He said later in describing the work of his Menlo Park days:

I do not regard myself as a pure scientist, as so many persons have insisted that I am. I do not search for the laws of nature, and have made no great discoveries of such laws. I do not study science as Newton and Kepler and Faraday and Henry studied it, simply for the purpose of learning truth. I am only a professional inventor. [The "only" was a nice touch.] My studies and experiments have been conducted entirely with the object of inventing that which will have commercial utility. I suppose I might be called a scientific inventor, as distinguished from a mechanical inventor, although really there is no distinction.

His aims, chosen with the memory of the unwanted vote recorder still niggling at the back of his mind, were those of the great research and development departments to be found a few decades later in most industrial corporations. In 1876 they were looked upon as revolutionary and drew criticism from two quarters in particular. The idea of applying science to industry brought down the scorn of many scientists and the idea of approaching the problems of industry scientifically aroused the suspicions of big business.

The industrial research laboratory which he planned to create in 1876 would have to be within easy distance of New York. Ideally, it would be situated in a small place without alternative attractions, the site of a community which would grow naturally around the self-contained team of specialists he hoped to employ. And it is not too fanciful to suggest that Edison, whose first concrete houses were designed decades later to alleviate the slum conditions of New York, had another requirement at the back of his mind. What he may well have wanted was a site where living conditions would be more agreeable, for himself and for his workers, than the grim streets of Newark.

Menlo Park, a small cluster of houses in rolling farmland country, on the Pennsylvania Railroad a few miles below Elizabeth and twenty-four miles from New York, was eventually discovered by Samuel Edison, the father who came from his home in Illinois to reconnoiter possibilities for his son. In 1876 the place was unknown to fame; within a few months Edison had given it a life of its own and within a decade had made it world-famous as the birthplace of the phonograph and the incandescent lamp.

He himself bought one of the larger available houses and from it supervised the building of the laboratory, a long, narrowish, two-story clapboard building whose shorter end carried an open balcony. Nearby stood the machine shop and a small carpenter's shop, while a number of other buildings, including a library, were added later. The whole area was surrounded by a rectangular fence of neat, white palings.

At this unpretentious site Edison now gathered an extraordinarily competent group of workers. From Newark he brought Charles Batchelor, John Kreusi, John Ott and a handful of his better workers. Later there was the mathematician Francis Upton from the College of New Jersey, eventually to become Princeton University; Ludwig Boehm the glassblower; and a score of others whom Edison would describe as "friends and co-workers."

This phrase gives a clue to the climate in which Edison achieved his Menlo Park successes. His enthusiasm, combined with his ability and boisterous spirits, kept all the staff in a perpetual state of high endeavor. There was no need to press a man to do his best; he did it automatically, and for as long as the friendly untidy boss, "the Old Man"—aged thirty—was doing the same. At Menlo Park, during the central and most creative part of his career, Edison evoked hero-worship of an almost tribal quality.

The center of this inventive company was the laboratory. Much of the ground floor was filled with tables for sensitive instruments, the tables themselves standing on brick foundations built deep into the floor and thus insulating the instruments from vibration. Separated from this section there was a chemical laboratory and also a chamber where experiments in comparing the intensities of different lights could be carried out.

The story above was filled with long benches on which there stood chemical and other scientific apparatus. Around the walls, shelves were filled with bottles of chemicals while at the end of the room, in a large glass case, there were samples of the world's precious metals as well as of the more costly chemicals. Books and journals, open at the page which had last been consulted, lay at random on the tables and, according to Francis Jehl, "when evening came on, and the last rays of the setting sun penetrated through the side windows, this hall looked like a veritable Faust laboratory." One item standing at the end of the room appeared almost grotesquely out of place—an organ, obviously used regularly. Its presence suggested that the routine of the laboratory might be less than orthodox, which was indeed the case.

Edison worked when the spirit moved him.

> He could go to sleep anywhere, any time, on anything [said one of his later colleagues]. I have seen him asleep on a work bench with his arm for a pillow; in a chair with his feet on his desk; on a cot with all his clothes on. I have seen him sleep for thirty-six hours at a stretch, interrupted for only an hour while he consumed a large steak, potatoes and pie, and smoked a cigar, and I have known him go to sleep standing on his feet.

Instead of the arm as pillow, he would sometimes use a chemical dictionary and so numerous were his ideas on waking that he was said to absorb them from the dictionary while asleep.

Meals were ordered as and when required and midnight suppers were frequent.

> It often happened [Jehl has recorded] that, while we were enjoying the cigars after our night supper, one of the boys would start up a tune on the organ, and we would all sing together, or one of the others would give a solo.
>
> Another of the boys had a voice that sounded like something between the ring of an old tomato can and a pewter jug. He had one song that he would sing while we roared with laughter. He was also great in imitating the tinfoil phonograph. When Boehm was in good humour he would play his zither now and then, and amuse us by singing pretty German songs.

From the first, visitors were welcome, and few more so than newspaper men, with whom Edison moved on terms of easy familiarity. Few left without a good story, and even the most hilarious proposals were discussed. Thus the suggestion that energy might be gained from cucumbers, since after all they absorbed sunlight, was developed by Edison in imaginative fashion. "If the normal process could be reversed," he said "we would obtain the sunbeams for practical use. The cucumbers would in that case be a storage battery for light and no doubt science would help the farmer raise a special kind for light production." Would he take it up himself? "Perhaps I shall later on," he joked. "First I want to show to the world that my electric light will do its work for a long while to come; there's no use burning up all your powder at once."

Some visitors appear to have been surprised at Edison's casual catering arrangements. Thus William Preece noted in his diary for 18 May 1877: "Another blazing day which I spent at a place called Menlo Park with Edison— an ingenious electrician—experimenting and examining apparatus. He gave me for dinner *Raw ham!* tea and iced water!!"

If visitors were critical of Edison, he could be critical of them, and quite outside the comparative privacy of a diary. Claiming to the *Washington Post*— in a perfect example of the pot calling the kettle black—that scientists were too often badly dressed, he said: "I remember Sir William Thomson when he came to see me had on a suit of clothes—I tell you. His trousers were too short for him; his coat was old and greasy, the collar came up above his ears, and his hat looked as if he had boiled soup in it. And that was his bang-up suit, too." The comment was made two months before the *New York Daily Graphic* described Edison as having "the characteristic features of an American, the nasal accent of a down-easter, and the slovenliness to be expected in a genius."

From Francis Jehl's account it appears that visitors considered "jolly and convivial" were especially welcome.

Some of the office employees would also drop in once in a while, and, as everybody present was always welcome for the midnight meal, we all enjoyed these gatherings. After a while, when we were ready to resume work, our visitors would intimate that they were going home to bed, but we fellows would stay up and work, and they would depart, generally singing some song like "Good Night Ladies."

High spirits made Menlo Park very different from most of the big industrial research and development organizations which were its successors. Size was of course one reason. Edison never employed more than a handful of workers and the close intimacy between them was as different from the atmosphere in a 1,000-strong organization as a naval destroyer's is from that in an aircraft carrier.

There was one other contrast. Today the complexity of technology, the number of disciplines involved in even the simplest of industrial innovations, make it less easy than in Edison's day for one man to provide the inventive, intellectual, and practical impetus. Today it is usually the team that has to provide the necessary head of steam. It is true that some of the problems which Edison set his workers were tackled with almost interdisciplinary techniques, yet he himself was in practice the main stimulator and the sole judge and arbiter of where the month's effort should be concentrated. Thus it was his ideas and his methods of working that dominated the whole enterprise.

The methods were a curious combination of the personally intuitive and the strictly scientific.

> I never think about a thing any longer than I want to. If I lose my interest in it, I turn to something else [he explained]. I always keep six or eight things going at once, and turn from one to the other as I feel like it. Very often I will work at a thing and get where I can't see anything more of it, and just put it aside and go at something else; and the first thing I know, the very idea I wanted will come to me. Then I drop the other and go back to it and work it out.

Letting the subconscious do the work was supplemented by one of two methods. When it came to chemical problems, Edison adopted the system of Luther Burbank, the plant breeder, who would sow an acre and then pick from thousands of plants the one which seemed the best from which to breed; but with problems of a mechanical nature he relied on one thing only: hard logical thinking.

Whatever the problem, whatever the method of tackling it, the attack on all fronts that Edison favored inevitably meant many failures. Far from being discouraged, he took the same view as Einstein. After failing in a year-long effort to work out a unified field theory uniting electrical and magnetic forces, the discoverer of relativity commented that no one would now have to waste time explaining that particular blind alley. In a similar way Edison consoled a colleague who was lamenting that thousands of experiments on a certain project had failed to discover anything. "I cheerily assured him," he said, "that we *had* learned something, for we had learned for a certainty that the thing couldn't be done that way, and that we would have to try some other way."

Trying some other way, concentrating on the problem until it was eventually solved, epitomized the spirit of Menlo Park, and in his reminiscences Francis Jehl quotes Edison's revealing criticism of one member of the staff:

> He knows a lot but he doesn't stick to the job. I set him at work developing details of a plan. But when he happens to note some phenomenon new to him, though

easily seen to be of no importance in this apparatus, he gets side-tracked, follows it up and loses time. *We can't be spending time that way!* We have got to keep working up things of commercial value—that is what this laboratory is for. We can't be like the old German professor who as long as he can get his black bread and beer is content to spend his whole life studying the fuzz on a bee!

Within little more than a year Edison had transformed Menlo Park. Batchelor and Upton also had private houses, Mrs. Jordan's boarding house was supplemented by three or four more, while fresh trade came to the small hotel from visitors to the laboratory. Within four years, as the laboratory became deeply involved in the incandescent lamp business, the number of employees rose to two hundred.

Over the new-style community, Edison, still only in his early thirties, ruled like a patriarch. His methods of hiring and firing and of running the business were both unusual. When Edward Acheson, later to become famous as the inventor of Carborundum, applied for a job he was taken to the main laboratory.

> At one of the tables sat three men; the center one in colored calico shirt, without coat, was introduced as Mr. Edison. The one on his left I knew afterwards to be Mr. William J. Hammer, and the one on the right as Mr. Francis R. Upton. Mr. Edison, placing one hand to his ear to indicate I should speak loudly, asked, "What do you wish?" I replied "Work." He replied, perhaps with impatience, "Go out to the machine shop and see Kreusi," and returned to the work absorbing his attention.

Edison's handling of routine business was just as casual, and Samuel Insull, who later became his secretary, abandoned his efforts to bring him into line.

> I never attempted to systematize Edison's business life [he admitted]. His method of work would upset the system of any office. He was just as likely to be at work in his laboratory at midnight as at midday. He cared not for the hours of the day or the days of the week. If he were exhausted he might more likely be asleep in the middle of the day than in the middle of the night, as most of his work in the way of inventions was done at night. I used to run his office on as close business methods as my experience admitted; and I would get at him whenever it suited his convenience.
>
> Sometimes he would not go over his mail for days at a time; but other times he would go regularly to his office in the morning. At other times my engagements used to be with him to go over his business affairs at Menlo Park at night, if I were occupied in New York during the day.

This happy-go-lucky way of life continued to characterize Edison's daily routine, if it could be called routine. Ideas continued to bubble up. He might be eating or talking with friends when something he saw, a topic of conversation, or an intruding memory, jogged up a technological possibility. Out from his pocket would come one of the 200-page yellow-leaved notebooks that after the move to Menlo Park superseded the former loose sheets. Down would go his thoughts, the date, and possibly one or more sketches. This might happen a dozen times a day and few notebooks lasted more than a week. Many lasted less and at his death Edison had filled 3,400.

The result was a stream of inventions more diverse than those which had come from the Ward Street works in Newark.

> By the simple inhabitants of the region [wrote one of his colleagues] he was regarded with a kind of uncanny fascination, somewhat similar to that inspired by Dr. Faustus of old, and no feat, however startling, would have been considered too great for his occult attainments. Had the skies overspreading Menlo Park been suddenly darkened by a flotilla of airships from the planet Mars, and had Edison been discovered in affectionate converse with a deputation of Martial [sic] scientists, the phenomenon would have been accepted as a proper concession to the scientist's genius.

Edison had been at Menlo Park little more than a year when he invented the device which, more than any other, can be claimed as his and his alone. In a long life he improved the stock ticker, made the telephone a practical proposition, perfected the incandescent lamp and worked out a system for the efficient distribution of electricity to light it. He helped to bring the moving picture into existence, and made the electric car a practical proposition by his design of an efficient long-life battery. Yet it is no criticism to stress that in all of these fields other men were working along lines parallel to his own and that his success depended, to lesser or greater extent, on the growing body of knowledge that the age was producing. In particular, it was Bell's telephone which Edison successfully revolutionized while on the production of the incandescent lamp—but not on its utilization—Swan and Edison reached the finishing tape neck and neck.

The phonograph, which later generations were to know as the graphophone, the gramophone, and the player, was in a different category. However, although Edison was without doubt the first man to build a working phonograph, he was not the first man to think of it. The birth of photography had induced more than one imaginative mind to ask whether what had been done for sight might not be done for sound. Thus Tom Hood had speculated in his *Comic Annual* in 1839: "In this century of inventions, when a self-acting drawing-paper has been discovered for copying visible objects, who knows but that some future Niepce, or Daguerre, or Herschel, or Fox Talbot, may find out some sort of

Boswellian writing paper to repeat whatever it hears?" Nearly forty years later came Charles Cros, a French amateur scientist whose conception of the phonograph overlapped Edison's development of the idea in much the same way that Wallace's idea of evolution overlapped Darwin's massive exposition in *On the Origin of Species.*

In April 1877 Cros wrote a paper in which he described a method by which the human voice moved a vibrating membrane; the membrane produced a tracing on lampblacked glass which was photoengraved on to a metal disk; and another membrane, moving over the rugosities of the photoengraved trace, reproduced the original sounds. Cros could find no financial support but took what was in those days a common step: he deposited his paper, unopened, with the Académie des Sciences in Paris. Some months later he appears to have discussed his project with the Abbé Lenoir, a writer of popular science articles, and on 10 October a description of Cros's apparatus, christened the phonograph, appeared in *La Semaine du Clergé* under Lenoir's name. Two months afterward, possibly stimulated by reports of Edison's work seeping back from across the Atlantic, Cros decided that his paper should be publicly read.

Meanwhile Edison had been bringing another idea from conception to birth. In 1876 he had taken out a patent for a telegraph repeater, a device which, in the words of the patent, was "a method of recording ordinary telegraphic signals, by a chisel-shaped stylus, indenting a sheet of paper, enveloping a cylinder or plate, along the line of the groove cut in the surface of the latter." The following year he was also experimenting with the telephone and his mind, he recalled,

> was filled with theories of sound vibrations and their transmission by diaphragms. Naturally enough, the idea occurred to me: If the indentations on paper could be made to give forth again the click of the instrument, why could not the vibrations of a diaphragm be recorded and similarly reproduced? I rigged up an instrument hastily, and pulled a strip of paper through it, at the same time shouting "Halloo!" Then the paper was pulled through again, my friend Batchelor and I listening breathlessly. We heard a distinct sound, which strong imagination might have translated into the original "Halloo." That was enough to lead me to a further experiment. But Batchelor was sceptical, and bet me a barrel of apples that I couldn't make the thing go.

The date of the incident is not established but a sheet of paper in Edison's records, bearing the date 18 July 1877, carries a rough sketch beside which there is the note:

> X is a rubber membrane connected to the central diaphragm at the edge, being near or between the lips in the act of opening it gets a vibration which is communi-

cated to the central diaphragm and then in turn sets the outer diaphragm vibrating hence the hissing consonants are reinforced and made to set the diaphragm in motion — we have just tried an experiment similar to this one. [A note further down the sheet says:] Just tried experiment with a diaphragm having an embossing point and held against paraffin paper moving rapidly. The spkg vibrations are indented nicely & there is no doubt that I shall be able to store up and reproduce automatically at any future time the human voice perfectly.

The notes adequately dispose of the often-repeated story that the phonograph was conceived when Edison's finger was pricked by the vibrating needle of an experimental telegraph repeater. He frequently denied the story but it still lives on, often being attributed to Edison himself.

Throughout the late summer he continued work on an improved repeater. He made some tentative experiments aimed at producing an incandescent lamp. And while he continued to turn over the possibilities of automatically recording the human voice, the project appears to have been well down the list of priorities. The situation was to be dramatically changed in November. On the 3rd the *Scientific American* announced that a Dr. Rosapelly and Professor Marey had succeeded in graphically recording the movements of lips and of a portion of the palate as well as the vibrations of the larynx. Furthermore, there was the prophecy that electricity might eventually allow these movements to be transmitted to distant points.

It is not certain what was in the air. But the announcement gave E. H. Johnson, who had been sent to Edison with Little's automatic telegraph a few years previously and had joined his staff, the opportunity to reveal in the journal on 17 November that at Menlo Park there had been conceived "the highly bold and original idea of recording the human voice upon a strip of paper, from which at any subsequent time it might automatically be redelivered with all the vocal characteristics of the original speaker accurately reproduced."

The editor of the journal seized on the idea and in an editorial speculated on the future of such a contrivance with an enthusiasm that may well have made Edison change his order of priorities.

Will letter writing be a proceeding of the past? [it boldly asked].... Are we to have a new kind of books. There is no reason why the creations of our modern Ciceros should not be recorded and detachably bound so that we can run the indented slips through the machine and in the quiet of our apartments listen again, and as often as we will, to the eloquent words. Nor are we restricted to spoken words. Music may be crystallized as well.

In England, *Nature* repeated the story, announcing that Edison was trying "to make the telephone record the sound it transmits.... The invention suggests an ultimate possibility of recording a speech at a distance, verbatim, without the need of shorthand."

Whatever added encouragement was given to Edison by the *Scientific American* and *Nature* articles, he had already been considering the use of something more efficient than paper and during the first days of December handed a rough sketch of a machine to John Kreusi. The sketch today bears the date "Aug. 22 1877," and it was for long thought that the phonograph had progressed from idea to reality that summer. Recent research by Byron M. Vanderbilt of the American Chemical Society has shown, however, that the date was written on years afterward and that the entries in Charles Batchelor's Day Book for 4 December: "Kreusi made the phonograph today," and for 6 December: "Kreusi finished the phonograph today," undoubtedly date the invention correctly.

The sketch showed a metallic cylinder around which a helical groove had been cut from end to end. The cylinder, mounted on a shaft which could be rotated by a handle at one end, was free to move horizontally along the shaft. On either side of the cylinder there was arranged a diaphragm and from each of the two diaphragms there projected a needle which could be moved into the helical groove on the cylinder.

On a date that was presumably Thursday, 6 December Edison's workers assembled in the laboratory to consider the result of Kreusi's work. Most of them were sceptical. So, apparently, was Edison.

> I didn't have much faith that it would work, expecting that I might possibly hear a word or so that would give hope of a future for the idea. Kreusi, when he had nearly finished it, asked what it was for. I told him I was going to record talking, and then have the machine talk back. He thought it absurd.

However, a sheet of thin tinfoil was now wound round the cylinder, and the needle projecting from one of the diaphragms was adjusted so that it rested on the tinfoil covering the start of the helical groove. Edison began to turn the handle of the machine and, as he did so, spoke into the diaphragm.

Compared with the "What hath God wrought," which Samuel Morse had sent as his first message, or Bell's almost accidental "Come here, Watson," which showed that the telephone really worked, Edison's words were banal. "Mary had a little lamb," he recited as he turned the handle, "its fleece as white as snow...."

While the rest of the men looked on, he withdrew the needle from the recording diaphragm, brought the cylinder back to its starting point, then adjusted the needle of the other diaphragm until it rested on the tinfoil.

Once again he turned the handle. As he did so there came from the machine the faint but unmistakable voice of Edison: "Mary had a little lamb...."

Kreusi exclaimed, as he well might: "Gott in Himmel."

"I was never so taken aback in my life," Edison said. "Everybody was astonished. I was always afraid of things that worked the first time. Long experience proved that there were great drawbacks found generally before they could be got commercial; but here was something there was no doubt of."

It was another eighteen days before he applied for a patent, during which time he and the staff improved the machine in many minor ways. Yet Edison would not have been Edison had he waited to give the world the news. Before the week was out he visited the offices of the *Scientific American* in New York, as the editor, Mr. Beach, announced in his columns.

> Mr. Thomas A. Edison recently came into this office, placed a little machine on our desk, turned a crank, and the machine enquired as to our health, asked how we liked the phonograph, informed us that *it* was very well, and bid us a cordial good night. These remarks were not only perfectly audible to ourselves, but to a dozen or more persons gathered around, and they were produced by the aid of no other mechanism than the simple little contrivance explained and illustrated below.

Edison's patent application, applied for in Washington on 24 December, and granted as No. 200,251 on 19 February 1878, covered a method of

> arranging a plate, diaphragm, or other flexible body capable of being vibrated by the human voice or other sounds, in conjunction with a material capable of registering the movements of such vibrating body by embossing or indenting or altering such material, in such a manner that such register marks will be sufficient to cause a second vibrating plate or body to be set in motion by them and thus reproduce the motions of the first vibrating body.

In addition it mentioned the use of a revolving disk rather than cylinder, a plaster-of-Paris process by which copies of an original record could be made, and various other possible improvements.

During the first few months of 1878 a number of improvements were incorporated in the succession of machines made at Menlo Park. It was found that one diaphragm could be used both to record and reproduce, while the indentations made, and the corresponding strength of the reproduction, could be increased by using a funnel-shaped attachment to concentrate the sound waves on to the diaphragm. Yet the cylinder still had to be rotated by hand. There was some difficulty in maintaining a constant speed as a message was being recorded, while in reproduction the cylinder had to be turned at exactly the same speed if the pitch was to be accurate. The problem was soon to be overcome by the addition of a motor while one visitor reported seeing a phonograph "run by steam power, with a belt through the floor to the machine shop." More awkward was the fact that a cylinder carrying a message could be replayed only a few times before the tiny indentations became so worn that the sounds were meaningless.

In Menlo Park, Edison delightedly played host to a constant stream of newspaper reporters. To the *New York Graphic* he gave his much-quoted comment: "I've made a good many machines, but this is my baby, and I expect it to grow up to be a big feller, and support me in my old age."

The *Graphic* reporter was among visitors given what was to become a typical demonstration by Edison. "Shall we give you a song?" he started.

as the suggestion is received with applause by the ten or a dozen visitors, he calls an assistant, adjusts a queer-looking double mouthpiece with two tubes meeting in one, and they sing, "Tramp, tramp, tramp, the boys are marching," the assistant singing tenor and Edison singing the air in a bass voice. He turns back the cylinder, puts on steam, and presently the machine begins to sing that famous war song, all the words being clearly articulated, and the two parts being perfectly distinct.

"What'll you have next?" asks Edison. "Shakespeare" replies the *Graphic*, and the inventor gives this to the machine in a solemn tone:

> Now is the winter of our discontent
> Made glorious summer by this sun of York.

With these first visitors, Edison was free both with his plans for improvements in the machine and with his ideas of its social impact. To start with he would use the sapphire point instead of the needle.

Two other things I am going to do [he went on] substitute some sort of membrane for this ferrotype-tympanum, and put some sort of a voice chamber over the mouthpiece about the size of the human mouth, with teeth and perhaps tongue. This will give the resonance that is lacking in this machine. Another thing I am going to do right away—abolish this whole cylinder and supersede it with a flat circular steel plate about as big as a dinner plate. This plate will be reamed with a fine groove running around itself, beginning in the center and ending in the circumference. I can make that groove so fine that the plate will hold 50,000 words, that is it will hold the whole of one of Dickens's novels. My puzzle now is to decide whether to make it fine enough to hold 50,000 words or coarse enough to hold only 200. A merchant may not want all his business mixed up on his phonograph. He may prefer to shift the tinfoil....

An entire Dickens novel could be printed on a sheet ten inches square which could be produced by the million. "These sheets will be sold for, say twenty-five cents," he explained. "A man is tired and his wife's eyes are failing, and so they sit around a table and hear ... read from this sheet the whole novel with all the expression of a first-class reader." A publications office would be set up in New York, he told the *World*, and would include sheets of novels, music and general literature.

> You can take a sheet from the album, place it on the phonograph and have a symphony performed. Then by changing the sheet you can listen to a chapter or two from a favorite novel and this may be followed by a song, a duet or a quartet. At the close the young people may indulge in a waltz, in which all may join, for no one need be asked to play the dance music.

These were heady ideas but they were not to be implemented for more than a decade. The phonograph was at this stage of its development no more than a novelty and it was as a novelty, undeveloped into anything better, that Edison began to exploit it. Before the end of January 1878, he had formed the Edison Speaking Phonograph Company and sold his rights in the machine to it for $10,000 cash plus a substantial royalty. The firm made and licensed the machines on an area basis to showmen who operated in fairground style. Throughout the United States the phonograph filled halls and drew crowds who listened with rapt attention not only to their own voices but to the music, the foreign languages, the lowing of cattle, the barking of dogs, and anything else which an ingenious showman could record.

Some demonstrations were lively affairs. In New York Edison himself took charge of one at which he had hired Jules Levy, a well-known cornetist. First, "Yankee Doodle" and other tunes were recorded and played back. After that,

> Mr. Edison showed the effect of turning the cylinder at different degrees of speed, and then the phonograph proceeded utterly to rout Mr. Levy by playing his tunes in pitches and octaves of astonishing variety. It was interesting to observe the total indifference of the phonograph to the pitch of the note it began upon with regard to the pitch of the note with which it was to end. Gravely singing the tune correctly for half a dozen notes, it would suddenly soar into regions too painfully high for the cornet even by any chance to follow it. Then it delivered the variations of "Yankee Doodle" with a celerity no human fingering of the cornet could rival, interspersing new notes, which it seemed probably were neither on the cornet nor any other instrument—fortunately. Finally the phonograph recited "Bingen on the Rhine" after its inventor, then repeated the poem with a whistling accompaniment, then in conjunction with two songs and a speech, all this on one tinfoil, though by this time the remarks began to get mixed. Just here Levy returned to the charge, and played his cornet fiercely upon the much-indented strip. But the phonograph was equal to any attempts to take unfair advantage of it, and it repeated its songs, and whistles, and speeches, with the cornet music heard so clearly over all, that

its victory was unanimously conceded, and amid hilarious crowing from the triumphant cylinder the cornet was ignominiously shut up in its box.

Edison loved it all!

If he had become "The Wizard of Menlo Park" and "the New Jersey Columbus" almost overnight, it was not entirely due to a fame that spread through what had become an offshoot of the entertainment industry. Another reason for the extraordinary public interest sprang from the suggestion which Beach had made in his *Scientific American* article—"the startling possibility of the voices of the dead being reheard through this device." Had Edison lived a generation earlier men might still hear the orders given by the Civil War commanders or Lincoln at Gettysburg. In the future, Beach went on, the voices of the great singers could be heard for

as long as the metal in which they may be embodied will last. The witness in court will find his own testimony repeated by a machine confronting him on cross-examination—the testator will repeat his last will and testament into the machine so that it will be reproduced in a way that will leave no question as to his devising capacity or sanity.

Edison himself had rather different views. Although the phonograph was to be exploited during the next few years largely for entertainment purposes, he saw it predominantly as a business aid. This he made clear in a long article in *The North American Review* where he flatly stated: "The main utility of the phonograph, however, being for the purpose of letter writing and other forms of dictation, the design is made with a view to its utility for that purpose." Two sheets of tinfoil could be indented at the same time, thus enabling the businessman to keep, as it were, a carbon copy of what he had dictated. Moreover he would now be able to record his messages rather than dictate them, and then put them in the post, and "*thereby dispense with the clerk*, and *maintain perfect privacy* in their communications."

Edison also forecast the use of his instrument for educational purposes as well as to keep, with the help of the telephone, a record of what a man had actually said. There were other potential developments, most of them to be exploited before the end of the century. But it is significant that the ever practical Edison gave priority to the stern utilitarian use of helping businessmen run their businesses more efficiently.

Meanwhile, the phonograph continued its progress as a nine-day wonder that lasted for weeks. There were still the skeptics. In France, where Edison's achievement was at first described in terms that gave him almost miraculous powers, it was soon being claimed as no more than a hoax involving the help

of a ventriloquist. And to Menlo Park there came an Anglican bishop who asked
if he could speak into a phonograph. Edison assented and the bishop then
recited, at rapid speed, a long string of tongue-twisting biblical names. Edison
moved the cylinder back to its starting point, turned the handle and watched
the amazement of his visitor as he listened to his own voice. The bishop confessed
he had suspected some kind of trickery. But there wasn't a man in the United
States who could recite the names at his speed.

The final accolade came in April 1878. Two months previously Sir William
Thomson had told an Edinburgh audience that the phonograph was "the most
interesting mechanical and scientific invention" that had been heard of in the
century. Now Edison was invited to demonstrate the phonograph before the
American Academy of Sciences. He arrived in Washington on the morning of
the 18th. For once immaculately dressed in a new check suit, and accompanied
by Charles Batchelor, he was met at the station by Uriah Painter, a well-known
newspaper correspondent who told him that Gail Hamilton, the society hostess
and niece of James G. Blaine, Speaker of the House of Representatives, had
asked if he would demonstrate the machine before members of Congress. What
is more, President Hayes had requested a demonstration at the White House.

To a *Washington Post* reporter Edison described the phonograph in words
which reveal his feelings:

> This tongueless, toothless instrument, without larynx or pharynx, dumb, voiceless
> matter, nevertheless utters your words, and centuries after you have crumbled to
> dust will repeat again and again to a generation that will never know you, every
> idle thought, every fond fancy, every vain word that you choose to whisper against
> this thin iron diaphragm.

He was first hurried off to the Smithsonian where he was greeted by Joseph
Henry. When all was ready, he began to turn the handle. "The speaking phonograph," announced the machine, "has the honor of presenting itself before the
American Academy of Sciences...." Edison followed with a brief description
of the working principles. Next, on to Gail Hamilton's salon to which, throughout the rest of the day and evening, there came a string of visitors, Congressmen,
city notables, and the influential inquisitive.

All went well except for one nasty moment. Among the visitors was Senator
Roscoe Conkling, a prominent member of Congress whose forehead was orna-
mented by a curl of hair that had become the cartoonists' delight but something
of an embarrassment to the Senator.

With the Senator present, Edison first recited into the machine the familiar
"Mary had a little lamb." At a loss for fresh verses toward the end of a long
day, he continued with another nursery rhyme:

> There was a little girl, who had a little curl
> Right in the middle of her forehead;
> And when she was good, she was very, very good,
> But when she was bad she was horrid.

The Senator was not amused and it was assumed that Edison had missed his name and failed to note the curl. But it was Conkling, counsel for Western Union in the multiplex case, who had grilled Edison in the witness box. The occasion no doubt offered a temptation that Edison could not resist.

It was eleven at night before word came that President Hayes would be glad to have the machine demonstrated at the White House. Here Edison was so successful that about half-past twelve, when it was expected he would leave, the President suggested that Mrs. Hayes and her friends should hear the talking machine. They were aroused from their beds and only after another three hours was Edison allowed to leave, quite certain that the phonograph was merely an augury of things to come. Faced by the *Washington Post* reporter's remark that he was only thirty-one, he had already commented: "I am good for fifty, and I hope to astonish the world with things more wonderful than this." A demonstration before the President had not diminished his confidence.

A week later his lawyers filed in London a patent for "Recording and Reproducing Sounds." This, with its twelve pages of specification and sixty-seven sketches, covered numerous additions to the earlier patent, including the notion that the material on which the sounds were recorded could be "in the form of a disk, a sheet, an endless belt, a cylinder, a roller, or a belt or strip." In August 1878, the patent was granted in England where Edison's old critic, William Preece, the British Post Office engineer, demonstrated the phonograph to good effect before the Physical Society.

Despite all the acclaim, it was still basically, apart from its entertainment role, a scientific curiosity. It was a fascinating machine for explaining the operation of sound waves but it was still neither the great aid to the businessman that Edison hoped nor the great entertainer that it eventually became. All this lay in the future and with most men in control it would have been the very near future. But Edison quickly lost interest once an invention was over the hump and a chain of circumstances now drew him away from development of the phonograph not for a few months or for a year or two but for a whole decade.

The first of the circumstances was the chance to test another of his inventions, the tasimeter, during the total eclipse of the sun which was due on 29 July. The purpose of the tasimeter was to record minute changes of temperature, the heart of the instrument being a button of carbon, the material whose effectiveness in transmitting electricity according to the pressure put upon it Edison had already used in the telephone's carbon transmitter.

In the tasimeter a strip of some material extremely sensitive to heat was mounted behind a funnel which was pointed at the object whose heat was to be measured. The top end of the strip, frequently vulcanite, was securely held. The bottom end rested on a metal plate below which there lay the carbon button; below this there was another metal plate, the carbon and the two plates being connected into an electric circuit incorporating a sensitive galvanometer. In operation, the heat from the object at which the funnel was pointed caused the vulcanite to expand, the lower end of the vulcanite strip pressed more firmly on the carbon button, and the change in pressure was reflected in the change of current shown on the galvanometer. Various ways of calibrating the results could be used and Edison claimed that it was possible to measure a change of as little as a millionth of a degree Fahrenheit. What this meant in practice he described in a letter to W. F. Barrett of the Royal College of Science for Ireland in Dublin:

> By holding a lighted cigar several feet away I have thrown the light right off the scale, and by increasing the delicacy of the galvanometer the tasimeter may be made so sensitive that the heat from your body, while standing eight feet away from and in a line with the cone, will throw the light off the scale, and the radiance from a gas jet 100 feet away gives a sensible deflection.

Edison never patented the tasimeter, believing that its use should be free to everyone. A modified version, mounted below the keel of an ocean liner, could give advance warning of icebergs. Its uses in fire detection were obvious and there were others that he envisaged if the instrument were as satisfactory in working tests as it was in the laboratory. The eclipse of 1878, offering the chance of measuring the heat of the sun's corona, provided just such an opportunity.

There was rather more to it than that, however. Until very late in life Edison rarely took a holiday in the accepted sense of the word. After all, he enjoyed inventing, he enjoyed discussing new ideas with his colleagues, and he enjoyed taking off his jacket, and finding out for himself whether the ideas worked. Thus he was one of the lucky few who saw no need to divide life into "work" and "holiday." Why should he spend his time idling, looking at the sights of foreign countries, or indulging in this occupation called sport when there were still exciting challenges to be met? His superabundant energy, moreover, enabled him to carry on, year after year, with a ceaseless stream of restless activities that would have worn out most men.

But there were exceptions, and one came now. "The general excitement over my invention and exhibition of the phonograph out at old Menlo Park frustrated serious or continuous work for a time in any other direction," he recalled. "In fact, my health gave way under the strain and in July I broke away for a Western trip as far as California." The recollection of health "giving way" is perhaps a mild exaggeration. The eclipse was to be seen at its best from Rawlins, Wyoming, then on the frontier of settled America, and Edison relished the two-month break as an ideal combination of work and jaunt.

He was certainly tired after the excitements and exertions of the previous few months, and this helps to account for an unjustified outburst which now helped keep alive all the existing academic suspicions nourished by the fact that he was more entrepreneurial inventor than scientist.

Late in May 1878, the British-born scientist David Hughes announced the development of a microphone that utilized the properties of carbon that Edison was utilizing in his loud-speaking telephone receiver. Hughes was an acquaintance of William Preece, with whom Edison had remained on friendly terms since the visit to Menlo Park in May 1877, and to whom he had passed on information about his current work. Ignoring the examples of simultaneous but independent discovery with which science is littered, and apparently not appreciating the differences between Hughes's work and his own, Edison cabled Preece, saying: "I regard Hughes heat measure and direct impact telephone as an abuse of confidence. I sent you, and others, papers describing it, also in letters about trouble with expansion telephones. If you do not set it right I shall, with details."

It was impetuous but it kept the accusation private. However, without waiting for a reply, Edison then brought the matter into the open with a letter to the *New York Tribune*.

> I freely showed [Preece] the experiment I was then making, including the principle of the carbon telephone and the variability of conducting power in many substances under pressure [it said in describing the 1877 visit]. I made him my agent for the presentation of this telephone, and subsequently of the phonograph, in England, and kept him informed, by copies of publications and by private letters, of my leading experiments, as he always manifested a great desire to be the means of presenting my discoveries to the British public. I therefore regard the conduct of Mr. Preece in this matter as not merely a violation of my rights as an inventor, but as a gross infringement of the confidence obtained under the guise of friendship.

Preece replied soberly enough with the statement that Hughes's microphone was different from Edison's telephone, that it had been worked out by Hughes without any information from him and that the first he knew of the microphone was when it was shown to Thomas Huxley; Norman Lockyer, the editor of *Nature*; and himself. Edison had no doubt been the victim of coincidence, although this does only a little to explain, and nothing to condone, his wild accusation. But his quick, if unconsidered, reaction was an index of the relentlessness with which the technological race was being run. It was also, perhaps significantly, typical of the accusations which were later to be made against Edison himself.

The party which gathered shortly afterward in Wyoming was international, astronomers from a score of countries having traveled to the township of Rawlins in a special carriage provided by the railroad authorities. Correspondents from the main newspapers were in attendance, and the trip had the flavor of a gala, even if a scientific one. The atmosphere of the place where the world's most famous astronomers had gathered is suggested by Edison's account of his first night in the town.

> My roommate was Fox, the correspondent of the *New York Herald*. After we retired and were asleep a thundering knock on the door awakened us. Upon opening the door a tall handsome man with flowing hair dressed in Western style entered the room. His eyes were bloodshot, and he was somewhat inebriated. He introduced himself as "Texas Jack"—Joe Chromondo—and said he wanted to see Edison as he had read about me in the newspapers. Both Fox and I were rather scared, and didn't know what was to be the result of the interview. The landlord requested him not to make so much noise, and was thrown out into the hall. Jack explained that he had just come in with a party that had been hunting, and that he felt fine. He explained, also that he was the boss pistol shot of the West; that it was he who had taught the celebrated Doctor Carver how to shoot. Then suddenly pointing to a weather vane on the freight depot, he pulled out a Colt revolver and fired through the window, hitting the vane. The shot awakened all the people and they rushed in to see who was killed. It was only after I said I was tired and would see him in the morning that he left.

Edison's main problem was to find a good site for the tasimeter. The best had been taken by the astronomers and he was forced to use a small yard at the end of which there was a chicken house. His troubles were increased by a strong wind that developed into a gale as the time of the eclipse approached. Every gust rocked the dilapidated chicken house and the vibrations forced Edison constantly to adjust the tasimeter. Wires and ropes were used in an attempt to secure it firmly, but the instrument was still swaying when the eclipse began shortly after two o'clock. The moon slowly shut out the sun and soon after three the last light began to go.

The chickens, Edison noted, went to roost as darkness fell. With the wind still blowing he again tried to adjust the instrument. Almost at the last moment he succeeded—only to find that the tasimeter had been too sensitive. The heat from the sun's corona was ten times too strong to give a satisfactory reading.

The eclipse over, Edison left Rawlins with a few colleagues for a trip that took them 100 miles south into territory only recently settled. The first part of the journey was by Union Pacific Railroad, a corporation now controlled by Jay Gould from whom Edison had acquired a special pass allowing him to travel on the cowcatcher at the front of the train. He appears to have enjoyed himself and a few months later was writing to the British astronomer, Norman Lockyer, who had also been on the trip: "I hope you will come over here again (after you have become well smoked up in London). With several other deep and mighty intellects we will take to the mountains for a grand hunt."

He had in fact taken part in several hunting trips deep into country still at the mercy of raiding Indian bands—a month later a patrol of thirty troops was massacred at one of their camp sites. He also visited a number of gold mines and here could not be restrained from the business of inventing. The result was a method of estimating how much ore was present when gold had been struck.

> The ore [he told J. B. McClure a few months later] is surrounded by a bed or bank of conducting material. For instance, in the mines which I examined that material was clay. The quantity of clay is an indication of the quantity of ore. When ore is struck thousands are often expended in drilling for more, when in reality the vein is completely exhausted. The contrivance I suggested enables the miner to know whether or not the vein is exhausted. I simply make a ground connection and run a wire through a battery and instrument. Now, I take the other end of the wire down the shaft and connect it with the clay or other conducting material surrounding the ore. If the clay bank is extensive the connection is a good one, and the current of electricity flows freely; but if the clay bank is small in area a poor connection is formed. By adopting a unit of measurement the area can be told almost to the square foot.

It is doubtful whether Edison's crude method would have been of much use. Nevertheless, methods using electrical resistivity of the soil and involving complex measurements over wide areas are currently used in prospecting for metallic ores, so once again Edison was in advance of his time.

After a quick trip by railroad to California, Edison returned to Menlo Park at the end of the summer, greatly refreshed and anxious to get down to work once again. The phonograph might have seemed the obvious thing to concentrate on. As a novelty it had been a success; but Edison had from the first intended it to be something more and it would have been natural had he now devoted his energies to transforming it. Instead, he turned to the problem that had been intriguing the scientific world for a quarter of a century.

Chapter Five

The Birth of the Bulb

Edison's journey to Wyoming to watch the eclipse of July 1878 was to have reper-cussions that had nothing to do with the tasimeter. As the train had taken the party across the newly settled plains of the Middle West he had remarked on the long road hauls made by farmers taking their grain to market or to the nearest elevator. To the man who had worked the Detroit–Port Huron line less than two decades earlier, an extension of railroad branch lines seemed to offer the obvious answer; and if steam was too expensive, as he judged it would be after some initial back-of-an-envelope calculations, then might not electricity be conscripted to do the job? There were other ideas which attracted him during the trip and at one point, watching miners laboriously at work in the Yosemite Valley, he turned to his companion, Professor George Barker from the University of Pennsylvania, with the question: "Why can't the power from the river be transmitted to these men by electricity?"

The belief that electricity could and should be used to ease human burdens was constantly at the back of Edison's mind as he worked away during the next decade and more. He enjoyed the challenge. He liked making money, even though his reasons were not the usual ones. But a feeling for what electricity could contribute to living conditions was never far below the surface and in the future he was to give it an almost mystic gloss. Many years later, back from a tour of Switzerland, he observed: "Where waterpower and electric light had been developed, everyone seemed normally intelligent. Where these applications did not exist and the natives went to bed with the chickens, staying there till daylight, they were far less intelligent." Simplistic, of course, but typical of the man who applied civilized criteria to the problems of technological progress.

In Menlo Park during the late summer of 1878 he once again mulled over the potentialities of electricity. In the fall of 1877 and first weeks of 1878 he had made a number of tentative experiments in search of a practical incandescent bulb, but had abandoned them because, as he put it, "so many others are working in the field." That was no doubt true although there is no doubt that the invention of the phonograph would in any case have drawn him away from work on the electric light. Now, early in September he was persuaded by Professor Barker to visit William Wallace, a well-known maker of dynamos in Ansonia, Connecticut. Wallace was experimenting with an improved series of arc lights, and it seems likely that Barker guessed what Edison's reaction would be to the glaring, flaring arcs that were the only "burners" using electricity to make light. He was right. As they left the works Edison turned to Wallace with the remark: "I believe I can beat you making electric lights. I don't think you are working in the right direction."

It was a confident, not to say presumptuous assertion, and at first glance the man who had made it gave little suggestion that it would be justified. Edison still showed few outward signs of technological genius. Indeed, to the end of his life he managed to retain the appearance of the countryman come to town, slightly wide-eyed and wondering; only those who looked more closely noticed the steeliness of the gray-blue eyes. Already famous but only on the verge of his greatest triumphs, he still had the mop of chestnut hair which was such a striking feature in the photographs of his youth.

> His forehead [wrote a New York reporter] is round and moderately full, but not high; nose prominent; mouth large but possessing a pleasant expression; and his chin is of the purely executive type, square and prominent. His manner is modest and retiring, and exhibits a total lack of egotism or self-assurance, a quality as rare as it is marvelous in one whose name and reputation are world wide and his achievements among the most astounding of the age.

This was the man who was about to give the world the incandescent lamp, "the glow-bulb," or what the world today knows as the electric light.

It was still gas that ruled the lighting industry. Despite the monopoly, or more accurately because of it, there were two views about the gas companies. On the one hand they were accepted as the force responsible for moving many parts of the country out of the oil lamp era. On the other, they had a domination which was often resented. The *New York Herald* noted in April 1879, when the possibility of breaking the gas monopoly first began to be seriously considered:

> The accumulated wealth of the gas companies has enabled them in many instances to make the most audacious claims on public patience and to override the law conserving public and private rights. Public highways are torn up and disfigured without scruple, and the terror inspired by the meter man is universal, for his frown means the darkness of desolation.

Early in the fall of 1878 Edison began, in his habitually methodical way, to gather as much background information as he could.

> I ... bought all the transactions of the gas engineering societies, etc. all the back volumes of the gas journals [he recalled]. Having obtained all the data and investigated gas-jet distribution in New York by actual observations, I made up my mind that the problem of the subdivision of the electric current could be solved and made commercial.

This was an illustration of Edison's best-known adage. It appears in various forms, the shortest being his comment to an interviewer that "genius is one-percent inspiration and 99 percent perspiration." His reply to a question about genius from Samuel Insull, for some years his secretary, more accurately reflects what he really thought: "Well, about 99 percent of it is a knowledge of the things that will not work. The other one percent may be genius, but the only way that I know to accomplish anything is everlastingly to keep working with patient observation."

In the days when Edison was investigating electric light, the alternative to gas, and a very limited one, was the electric arc. Sir Humphry Davy had produced the first as far back as 1812, connecting the wires of a battery to two pieces of charcoal, separating the pieces and then watching the dazzling glare of the flame between them. The glare was one characteristic that restricted the use of the arc. Another was that the distance between the charcoal—or other forms of carbon which were soon substituted for it—had to be continually adjusted. A third disadvantage was the smell and smoke as the carbon points were oxidized, an apparently inevitable by-product of the process which virtually ruled out its use indoors. Notwithstanding these drawbacks the arc had from the 1850s had a successful if restricted use. An electric arc blazed out from the Clock Tower of London's Palace of Westminster whenever Parliament was sitting. An electric arc was installed in the lighthouse at the South Foreland on the English Channel while in America, Britain and France it slowly came into use for street lighting.

A year before Edison was persuaded to turn his inventive energies to the electric light an improvement in the arc had appeared as the Jablochkoff candle. This consisted of two molded carbon rods separated by an insulating layer of plaster of paris. When the candle was "lit," the insulating layer incandesced and burned away at the same rate as the carbon rods. Jablochkoff candles were used in 1877 to light part of the Avenue de l'Opéra in Paris and part of the Thames Embankment in London, but the largest practical size burned for only two hours and offered no real opposition to the gas burner.

The incandescent lamp appeared to be even less of a rival. It had been a possibility considered by scientists ever since Jobart had in 1838 heated a carbon rod, sealed inside a vacuum, and watched it glow as a current was passed through it. J. W. Starr, an American from Cincinnati, had patented an incandescent lamp in England as far back as 1845. Dr. J. W. Draper constructed a platinum lamp shortly afterward. W. E. Stairs did the same in 1850 and was followed by E. C. Shepherd. In 1858 Moses G. Farmer lit a room in his house at Salem, Massachusetts, for several months with platinum lamps controlled by automatic regulators, and a decade on, the Russian Lodyguine had lit up the Admiralty Dockyard at St. Petersburg with 200 lamps of his own design. They had been comparatively inefficient, and extremely expensive since it was necessary to renew them after less than twelve hours. In England, Joseph Swan had for years been experimenting with his own lamp while in the America of Edison's day other men tackling the problem included William E. Sawyer and Hiram Maxim, the extraordinary inventor of the Maxim gun and a steam-powered aircraft.

There was thus no shortage of men trying to break the near monopoly which gas exercised in the lighting industry. Yet all were faced with the immense gap between theory and practice. In theory, all that was needed was a thin filament, probably of carbon but possibly of some other material, sealed in a glass container from which the air had been driven; if a current were passed through the filament it would then glow to incandescence. The practice was not so straightforward. There was the problem of making a filament that did not break, then of producing a good enough vacuum. These were problems that had to be overcome before a satisfactory light bulb could be made even under laboratory conditions; if it came to commercial manufacture a mass of additional difficulties would have to be tackled.

Furthermore, as Edison was to stress, the circumstances were unusual.

> We have an almost infinitesimal filament heated to a degree which it is difficult for us to comprehend, and it is in a vacuum, under conditions of which we are wholly ignorant. You cannot use your eyes to help you in the investigation, and you really know nothing of what is going on in that tiny bulb. I speak without exaggeration when I say that I have constructed 3,000 different theories in connection with the electric light, each one of them reasonable and apparently likely to be true. Yet in two cases only did my experiments prove the truth of my theory.

The development of the incandescent lamp was a challenge apparently tailor-made for Thomas Alva Edison.

His aim was expressed simply enough in one of his notebooks under "Electricity vs. Gas as General Illuminants." "Object," it read, "Edison to effect exact imitation of all done by gas, so as to replace lighting by gas by lighting by electricity. To improve the illumination to such an extent as to meet all requirements of natural, artificial, and commercial conditions."

The first job was to discover the best material for the filament and the most effective shape into which to fashion it. In his experiments the previous year Edison had used carbonized paper, various tissues, wood, corn and other specially chosen fibers. Literally scores of different materials were tried, but with little success, partly due to the fragility of the filaments, partly to the lack of a high enough vacuum. Now, in the fall of 1878, he set to work once more.

At this point Edison was forced to look about for cash. He had enough experience to recognize that a large outlay of money on experiments would be inevitable and that even then success could not be guaranteed. But there was one field in which the expenditure of huge capital had already brought huge success and immense profits. That was telegraphy, and it is not surprising that of the twelve men who were persuaded to back Edison's incandescent lamp, eight had connections with the telegraph industry. The first to come to his aid was Grosvenor P. Lowrey, the general counsel of Western Union. Lowrey, who with Roscoe Conkling had cross-examined him without mercy in the multiplex case, suggested the setting up of a corporation to finance research and to take out patents. He quickly gained the support of Dr. Norvin Green, the President of Western Union; Tracy R. Edison, a leading stockholder in the Gold and Stock Telegraph Company; and Egisto P. Fabbri, a partner in J. P. Morgan.

The Edison Electric Light Company was floated with 3,000 hundred-dollar shares of which 2,500, plus $30,000 in cash, went to Edison. Its aims were made clear in the brief outlining what it planned to do. It had been organized first to own, then to license the use of, Edison's inventions in electricity, relating to lighting, power or heating—in fact those concerned with any use other than telegraphy. There was good reason for the brief being so comprehensive. Edison knew, better than any of his backers, that perfection of an incandescent bulb would not be the end of the road but only the beginning.

In his first experiments in the fall of 1878 he returned to his first love, carbon. "Around October and November, Batchelor made a great number of paper carbons, at least fifty, from tissue and other kinds of paper, coated over their surface with a mixture of lampblack and tar, rolled them up into the fine long form of a knitting needle, and then carbonized them," he wrote. "These we brought into circuit and brought up to incandescence *in vacuo*; although they would last but an hour or two. We tried a great many experiments with paper carbons, wood carbons, and some made from carbonized broom corn. What we desired at that date, and had settled our minds upon as the only possible solution of the subdivision of the electric light, was that the lamps must have a high resistance and small radiating surface."

A striking, and typical, picture of Edison during these months was drawn by William H. Bishop in an article in *Scribner's Monthly*. Entitled "A Night

With Edison," it was to have unexpected repercussions on Edison's future.

> It is much after midnight now. The machinery below has ceased to rumble and the tired hands have gone to their homes. A hasty lunch has been set up. We are at the spectroscope. Suddenly a telegraph instrument begins to click. The inventor strikes a grotesque attitude, a herring in one hand and a biscuit in the other, and with a voice a little muffled with a mouthful of both, translates aloud, slowly the sounds intelligible to him alone: "London May— News of death of Lord John Russell premature. John— Blanchard— whose— failure— was— announced— yesterday— has— suicided (no, that was a bad one, *succeeded*) in adjusting— his affairs,— and— will— continue— in business."

In spite of all his efforts, Edison's experiments with carbon showed no signs of success and he next turned to platinum and allied metals since these seemed likely to lead to the high resistance and the small radiating surface he considered essential. The need for these characteristics arose from his plans to "subdivide the electric light" as it was called, both by having smaller light sources than the arc and by "piping" electricity supplies around like gas and water. Current engineering opinion viewed this with skepticism. Yet from the first Edison had realized that some such "division" would be necessary. The unit of electricity in the arc light was too large for homes and small buildings; instead of one blazing arc there would have to be a number of less powerful bulbs. However, if the bulbs were coupled up in series—that is with the component parts of the electrical circuit connected end to end to form a single continuous path for the current—then turning off, or failure, of any one lamp in the circuit would break the path for the current and extinguish them all.

Quite apart from this problem there was the objection raised by W. H. Preece, the electrical engineer who comes up in the records as honest broker, criticizing Edison without inhibition when he thought fit and praising his ability, on different grounds, almost in the same breath. Writing in *The Telegraph Journal and Electrical Review* in February 1879, Preece pointed out:

> It is, however, easily shown (and that is by the application of perfectly definite and well-known scientific laws) that in a circuit where the electromotive force is constant, and we insert additional lamps, then when these lamps are joined up in one circuit, i.e. in series, the light varies inversely as the square of the number of lamps in the circuit and that, joined up in multiple arc, the light diminishes as the cube of the number inserted. Hence a subdivision of the electric light is an absolute *ignis fatuus*.

The skeptics were in virtually every case thinking of a fixed amount of current which would have to be divided among a number of lamps. Edison believed, however, that it would be possible to solve the problem by varying the current according to the number of lamps being fed; in other words by using a system in which the voltage would be fixed but in which the amperage would vary with demand. This meant that the dynamo used to power arc lights, and giving constant amperage, would have to be redesigned to give constant voltage. It also appeared to mean heavy losses of energy in transmission or the use of so much copper in the transmission system that its cost would make the system uneconomic. The problem was reduced by the choice of a voltage high enough to allow distribution over a practical distance but not so high as to destroy the flimsy filaments of the bulbs: the 110 volts which is still a common supply pressure in the United States. What remained of the problem was eventually removed by the feeder system and development of the three-wire system, conceived by Edison and others at about the same time.

However, the initial problem was the incandescent lamp. Of the metals tested in the fall of 1878 platinum at first seemed to provide the best answer and early in October Edison lodged a patent for "Electric Lights" which incorporated platinum filaments. The bulb—still known as burner, the term used in the gas industry—was a complicated affair compared with the bulbs which were in mass production less than two years later. Its filament, consisting of a double helix of platinum wire, incorporated a rod which expanded as the filament became hotter, created a short circuit before its temperature reached the melting point of platinum, and allowed the platinum to cool. The rod contracted and the current then flowed once again, an electrical example of the mechanical governor which had made George Watt's steam engine a practical proposition. Some months later Edison patented a similar device in which the make-and-break mechanism was operated by an expanding diaphragm.

The first patent for an incandescent lamp gave him a whiff of success and he confidently announced: "I have just solved the problem of the subdivision of electricity indefinitely." To the *New York Sun*, moreover, he gave a colorful description of the world that was just round the corner. Public and private buildings throughout New York would be lit by electricity using the power from about twenty separate stations; as for distribution, that would simply be a matter of passing electric wires through the existing gaspipes, which would no longer be needed. "All that will be necessary," he said "will be to remove the gas burners and substitute electric burners. The light can be regulated by a screw the same as gas." Edison was not particularly worried about how he would know the amount of electricity used by each consumer. "I know that it can be measured, but it may take some time to find out how," was his explanation. "I propose that a man pay so much for so many burners whether he uses them or not. If I find that this works an injustice, why I shall try to get up a meter, but I fear it will be very hard to do it." Town lighting was, in any case, only a part of his plans. Generators, he said, could be installed in steamboats or on trains, while in small country towns, water could provide the power.

All this was true, and the ignoring of practical difficulties that still remained was merely a result of the superabundant confidence that Edison usually exuded and that he expected of others. Years later, after listening to an employee saying that an important problem was almost solved, he commented: "He hasn't got a damn thing. But that's the way to talk."

In 1878 many obstacles still had to be overcome and it was not long before Edison dropped the use of platinum. However, the first patent was of considerable importance since it described the essential characteristics of the incandescent lamp as it was finally perfected and was thus of use in the spate of lawsuits which followed during the 1880s.

Despite his optimistic announcement, Edison was well aware of the hurdles ahead and he now redoubled his efforts. He realized that his own mathematical training was slight and that ultimate success with the incandescent bulb would be followed by a good deal of complex calculation if it were to be properly exploited as the vital part of a complete system. Before the end of the year, therefore, he had recruited Francis R. Upton, who had studied under Hermann Ludwig Ferdinand von Helmholtz in Germany and who was to prove invaluable in the months ahead. One of Upton's first findings confirmed what Edison had felt intuitively: that the lamp must have at least 100 ohms resistance to compete successfully with gas.

Yet however much help could be given by Upton, a great deal of the work still had to be carried out by a process of trial and error. The bulbs themselves could be circular, oval or columnar, and each shape was experimented with while enclosing a succession of different filaments. The exact volume of the bulbs was important in estimating their efficiency, a fact which was the starting point for one of the most repeated, though possibly apocryphal, Edison stories.

A year or two later when, after the initial success of the incandescent bulb, Edison was still experimenting with fresh shapes, he was found in the Menlo Park library by his secretary who "noticed that he had quite a twinkle in his eye." Two mathematicians on the staff were busy at work on some calculations and Edison said that they were working out the cubic contents of a certain bulb. "You come back here about six o'clock tomorrow morning, and they will still be figuring," he said.

I went home and went to bed and left a call for about half-past five [the secretary recalled]. I was up in the library again at six, and Mr. Edison told me to go over to the laboratory. The mathematicians were still figuring in the library, and they had been at it all night. I was told to get an incandescent lamp bulb that had not been used, one of the standard size, and to fill it with water; that I would also find in the same room a graduated glass, such a glass as druggists use, and then to bring the lamp bulb full of water and the graduated glass back to the library.

Edison took the bulb full of water in one hand and the graduated glass in the other, and poured the contents of the incandescent lamp bulb into the graduated glass. Of course there were marks on the graduated glass, and he knew exactly the contents of the incandescent lamp bulb on which the mathematicians had been figuring all night without coming anywhere near it.

After the lodging of the first incandescent bulb patent he redoubled his efforts, working even longer hours, immersed in his experiments to the exclusion of everything else, and as unthinking about danger as he was for the whole of his life. Thus in his notebook for 28 January, he recorded:

Owing to the enormous power of the light (alloy of nickel) my eye commenced to pain after seven hours work and I had to quit. Suffered the pains of hell with my eye last night from 10:00 P.M. to 4:00 A.M. when I got to sleep with a big dose of morphine. Eyes better, do not pain much at 4:00 P.M. but I doze today.

While the work went on, so did the regular flow of good news about the progress of research, newspapers tending to encourage the more colorful reports. This did Edison little good in the more staid scientific world. However, in some ways he was not a free man. His backers wanted results. They sometimes forced him into giving demonstrations before sufficient tests had been completed and, while he was still experimenting with metallic filaments, a number of them visited the laboratories.

It was dark when they appeared in the machine shop where the demonstration was to take place. The dynamo was started up.

Today I can see these lamps rising to a cherry red, like glowbugs, and hear Mr. Edison saying "a little more juice," and the lamps began to glow [Francis Jehl has written]. "A little more," is the command again, and then one of the lamps emits for an instant a light like a star in the distance, after which there is an eruption and a puff; and the machine shop is in total darkness. We knew instantly which lamp had failed, and Batchelor replaced that by a good one, having a few in reserve near by. The operation was repeated two or three times with about the same results, after which the party went into the library until it was time to catch the train for New York.

That demonstration virtually marked an end to work with metallic filaments.

While still working with platinum, however, Edison had quite accidentally made what was to be a significant discovery. Experimenting with a sliver of the metal inside a mechanical pump from which the air had been exhausted, he found that the pump appeared to be leaking. Batchelor, asked for an opinion, thought the same. Yet the pump, it became evident, was not in fact leaking. The small traces of gas—the occluded gases—appearing were found to have been contained in the metal and released when the metal was heated in a vacuum. Once that had been done, the metal could be heated still further without reducing the vacuum, that all-important requisite for the incandescent bulb.

Meanwhile, Edison worked on relentlessly, experimenting with material after material, producing filaments of different thicknesses and in different shapes. Some were too fragile to be satisfactorily sealed into the glass bulbs. Others burned out almost as soon as the current was passed through them. More complicated methods were still being tried and one lamp appears to have incorporated a built-in rheostat, the light being switched up or down by turning a small metal wheel beneath the lamp itself.

> By turning [it] all the way round [wrote one visitor] the full power is obtained; by turning only half-way round only half the full light is given. In this way the current can be regulated to any degree of light, from a dull red giving only the faintest glimmer, to a beautiful brilliant white-shedding light in every corner of the largest room.

As he kept doggedly on, receipt of a message from his agent in England triggered off an entirely contrary quality, his ability to change scientific horses in midstream if circumstances called for it. The message contained Colonel Gouraud's news about the possibility of action by the Bell telephone company. The "loud-speaking telephone" soon dealt with that, and within a few weeks the Menlo Park taskforce, having been switched to telephony, was switched back to the incandescent lamp.

By this time, the fall of 1879, Edison had obtained one of the new Sprengel air pumps which enabled him to produce a vacuum of about one hundred-thousandth of an atmosphere. He soon found that hermetically sealed bulbs would retain this high vacuum, certainly for long periods and possibly indefinitely. The same was true when, within a few weeks, he had adapted the pump to get a vacuum of a millionth of an atmosphere, a development which dramatically altered the picture. Edison had always felt that carbon was ideal for a filament and had been diverted from it only by the fact that carbon filaments burned out too quickly in the low vacuum previously possible. Now he turned back to carbon once more, using as his raw material a variety of threads, fibers, and similar substances.

The breakthrough came at the end of a five-day session beginning on 16 October. Edison was now using cotton thread, fitted into a hairpin-shaped groove cut in a nickel plate and then cooked and carbonized for a number of hours. But this was only the first stage in a long process, as he explained to Francis Arthur Jones.

All night Batchelor, my assistant, worked beside me. The next day and the next night again, and at the end of that time we had produced one carbon out of an entire spool of Clark's thread. Having made it, it was necessary to take it to the glassblower's house. With the utmost precaution Batchelor took up the precious carbon, and I marched after him, as if guarding a mighty treasure. To our consternation, just as we reached the glassblower's bench the wretched carbon broke. We turned back to the main laboratory and set to work again. It was late in the afternoon before we had produced another carbon, which was again broken by a jeweler's screwdriver falling against it. But we turned back again, and before night the carbon was completed and inserted in the lamp. The bulb was exhausted of air and sealed, the current turned on, and the sight we had so long desired to see met our eyes.

The current was increased. The bulb still burned. It burned for many hours, and a second one burned for forty. The experiments of 21 and 22 October proved that Edison was over the hump. "I think we've got it," he said. "If it can burn forty hours, I can make it last a hundred."

Within a fortnight he had applied for a patent covering the carbon-filament lamp, asserting that "even a cotton thread, properly carbonized and placed in sealed glass bulbs, exhausted to one-millionth of an atmosphere, offers from 100 to 500 ohms' resistance to the passage of the current and that it is absolutely stable at a very high temperature."

During the next few weeks everyone at Menlo Park concentrated on the task of making more bulbs. Not all were alike since Edison continued with his trial-and-error experiments, altering the size and shape of the glass container and the methods by which the wires were led into it and sealed. Meanwhile his colleagues turned out replicas of the prototype. Some were hung up in the laboratory itself. Others were taken to Edison's own home, to Mr. Batchelor's, and to Mrs. Jordan's boarding house, then connected to the generator at the laboratory by wires strung on poles. Yet others were suspended above the streets, and burned brightly through the night, attracting all who could walk or ride in from the surrounding countryside to gaze at the new wonder.

Edison had for long been adept at getting publicity for his activities, the result not merely of sound business instinct but also of the enthusiasm which bubbled out and affected even the most hardbitten journalist. Just as he was one of the first to recognize that success in the growing technological industries rested heavily on cooperative research, so was he one of the first to realize that advertising paid—particularly when it was free. "Edison was not seeking publicity," Francis Jehl has ingenuously and revealingly written of this period, "but often the newsmen from New York or elsewhere were assigned to Menlo Park for an 'interesting' story, and Edison always enjoyed helping them." Much that was written about his work was the result of inspired leaks, sprung either by Edison himself or by his colleagues; some came by imposing a pseudo secrecy which virtually compelled the press to go on investigating and rarely failed to produce results.

Now he realized that something different was needed. The lights of Menlo Park were already creating not only wonder but rumour and counter-rumour, and Edison appreciated it was necessary for some full and authoritative account to be given the widest possible circulation. Marshall Fox of the *New York Herald*, his rooming companion on the Rawlins eclipse expedition the previous year, was therefore invited to the laboratory, given a free hand to go anywhere and ask whatever questions he wished. In return, he guaranteed to hold the article he would write—which was almost certainly vetted by Edison himself—until publication was authorized from Menlo Park.

Fox came, and wrote. Then, through what appears to have been a genuine mistake, he was given premature permission to publish. The article appeared on 21 December. It covered an entire page of the newspaper, plus an additional column, and its importance can be judged from the eleven headings and sub-headings above it: "Edison's Light," "The Great Inventor's Triumph in Electric Illumination," "A Scrap of Paper," "It Makes a Light, Without Gas or Flame, Cheaper Than Oil," "Transformed in the Furnace," "Complete Details of the Perfected Carbon Lamp," "Fifteen Months of Toil," "Story of His Tireless Experiments with Lamps, Burners and Generators," "Success in a Cotton Thread," "The Wizard's Byplay, with Bodily Pain, and Gold 'Tailings,'" "History of Electric Lighting."

The article lived up to its promise, explaining with considerable accuracy how Edison's light was produced "from a little bit of paper—a tiny strip of paper that a breath would blow away." The light itself was "like the mellow sunset of an Italian autumn," while the bulb was "a light that is a little globe of sunshine, a veritable Aladdin's lamp." Edison always spoke of the article with considerable respect although its language did nothing to damp down the criticism that he was making claims he would not be able to support. More disturbing, it revealed the essentials of his work to competitors who were able to digest them while he himself was still finalizing his plans. However, on the credit side, it transformed the impromptu Menlo Park exhibition from a comparatively local display into a magnet whose lines of attraction spread out across the far wider area served by the *Herald*.

Two days after Christmas the New York correspondent of the London *Times*, the President of the Philadelphia Local Telegraph Company, and an assistant editor of the *Philadelphia Public Ledger*, arrived in a party at Menlo Park. From the report in *The Times* it seems that they were open-minded but skeptical.

Up to the present winter all the efforts of Mr. Edison have been dead failures. All the startling announcements made on his behalf have been premature. None of them have been justified by the facts. They indicated what the inventor hoped to accomplish, not what he had actually accomplished; and as, latterly, it had gradually been making itself known that Mr. Edison had failed, there was beginning to be great scepticism in the public mind as to his ever accomplishing anything important in this field of experiment.

The three men arrived about noon. They spent the next few hours touring the laboratory and receiving Edison's answers to their questions. All the time some of the new incandescent lamps were within sight, burning brightly, but it was only as darkness fell that the party fully appreciated what had been accomplished.

We ate our supper by [electric light] in the little restaurant that had been established at the Park, and I sat down in Edison's office under two of his lamps attached to a gas bracket and wrote the rough draft of the telegram sent to *The Times* [its correspondent said]. In this room a telegraph operator worked in a corner with an Edison lamp in a movable table stand illuminating his work. Downstairs his bookkeeper was paying off the hands by the aid of two more electric lights on a gas bracket. Out in the roadway in front of the building two street lamps were set up with the Edison lamp in full operation.

In Edison's home, a few hundred yards away, the visitors saw the revolution for life indoors that the incandescent lamp would bring about. "I shall never forget the glee with which Edison listened to the reading of a newspaper slip wherein an ambitious 'expert' offered to forfeit $100 for every lamp that Edison could keep burning for over twenty minutes," the *Times* man went on. Told that most of the lamps had been burning for three days, he and his colleagues watched for the obligatory twenty minutes before leaving.

Quick to exploit the situation, Edison had announced that the laboratory would be open to the public after Christmas. The Pennsylvania Railroad now ran special trains to Menlo Park and within a few days Edison was coping with a public spectacle.

For more than a week now [reported the *Herald* early in January 1880] the entire establishment with its twenty or thirty skilled hands, has been practically at a standstill, owing to the throngs of visitors. They come from near and far, the towns for miles around sending them in vehicles of all kinds—farmers, mechanics, laborers, boys, girls, men and women—and the trains depositing their loads of bankers, brokers, capitalists, sightseers, hungry agents looking for business.

It was one of the local visitors, an elderly man, who gave the filaments a nickname that was to pass into the language. "It's a pretty fair sight," he said, "but danged if I see how ye git the red-hot hairpins in the bottles."

Many visitors, having walked up the narrow plank road from the station to the laboratory, clamored to have a word with the great man. Others, ignoring the notices warning sightseers not to touch the apparatus, began experimenting on their own and one of the expensive vacuum pumps was among the equipment wrecked. By 3 January, fourteen lamps had been stolen and Edison was at last forced to announce that in future only visitors with special permits would be admitted to the laboratory.

Apart from its attendant troubles, the Menlo Park display was a triumph which a less flamboyant exhibition would have failed to achieve. It proved to thousands of ordinary people that the incandescent lamp really did work and it whetted the public appetite for a method of lighting that obviously had immense advantages over both gas and the arc light.

There remained, of course, the question of cost. Edison had no doubt about the answer to that: "After the electric light goes into general use," he told one visitor, "none but the extravagant will burn tallow candles."

Chapter Six

Teething
Troubles

Edison's spectacular demonstration at Menlo Park in the last week of 1879 confirmed him in the belief that he was leading a new industrial revolution. But the core of his idea had always been the distribution of electricity over a wide area from a central generating station. The problems presented were far greater than those of lighting one small group of buildings and he knew that the incandescent bulb itself was but one item, if the essential one, in a new and complex system which still had to be designed and constructed.

Reaction to the dramatic display was varied and contrasted. Neither the thousands of ordinary people who had thronged the streets of Menlo Park, gazing in awe and hope at the new lights, nor the newspaper correspondents, felt inclined to deny the evidence of their own eyes: problems might lie ahead, but the incandescent electric light had certainly arrived. "Informed" opinion knew better.

Elihu Thomson, the electrical engineer, saw little chance of the incandescent lamp system being profitable when he visited Menlo Park early in 1880 while even a year later the eminent Werner von Siemens believed that it could not compete economically with the arc lamp. On 5 January, moreover, the *Herald* published a revealing letter from W. E. Sawyer, who had so far failed to produce an incandescent lamp and who could not believe that anyone else had succeeded.

> Notwithstanding the assertion that one of Mr. Edison's electric lamps has been running for 240 hours, I still assert, and am prepared to back up my assertion, that Mr. Edison cannot run one of his lamps up to the light of a single gas jet (to be more definite, let us call it 12 candlepower) for more than three hours.

The *Herald* proposed in a leading article that Sawyer should see for himself at Menlo Park, adding "perhaps Edison would keep up the illumination many hours extra for the sake of converting this doubting Thomas."

The Sawyer letter was one of the early shots in what was to be a continuous barrage. The *Herald's* leading article was percipient enough at this early stage to suggest how its readers should react to them.

Mr. Edison will probably be the subject of many such letters before his system is in practical use throughout any large city. The best thing he can do is to burn them. A man who has on his hands such a gigantic enterprise as the electric light is claimed to be is under no obligation to waste a moment in giving points to a rival. Letters and criticism will be written for all sorts of motives—some of them honest, others to help the gas stock market for a day or to bring desired capital to the coffers of some rival, or even to bear Edison stock, so that the writers may buy some at a decline; still others will write mistakenly, because, although they never thought of the possibility of such a thing, they do not know all or exactly the same things that Edison does. Let the inventor stick to his laboratory, making his tests to the public only; then, if anyone chooses to remain unconvinced, in spite of his eyes, he will not be missed from the crowd.

Yet if Sawyer was still unbelieving there were others who not only glimpsed the significance of Edison's achievement but appreciated that he was on the brink of fame, fortune and the creation of a huge new industry. This prospect naturally enough aroused other inventors who had been struggling—in some cases for far longer than Edison—to make an incandescent lamp. The most important was the Englishman, Joseph Swan, who as far back as 1860 had been working along very similar lines. In the 1860s it was impossible to create the high vacuum that Edison had utilized, but by 1878, in fact within a few weeks of Edison's application for his first incandescent lamp patent, Swan was demonstrating a comparable lamp before the Newcastle Chemical Society. On this occasion it gave light for only a few minutes before being burned out by an unexpected surge of current. Nevertheless, Swan in England and Edison in the United States were achieving success at virtually the same time; and it was soon obvious that victory in the race for a practical incandescent lighting system rested largely on the development of a viable distribution system and the mass of auxiliary equipment which this involved.

It was in this situation that Swan had read the enthusiastic, and at times extravagant, claims coming from America. By the end of 1879 he could keep quiet no longer and on 1 January 1880, the day after the excited crowds had been walking the streets of Menlo Park, a letter appeared over Swan's name in *Nature*.

Fifteen years ago [this said] I used charred paper and card in the construction of an electric lamp on the incandescent principle. I used it too in the shape of a horse-shoe precisely as you say Mr. Edison is now using it. I did not then succeed in obtaining the durability which I was in search of, but I have since made many experiments on the subject and within the last six months I have, I believe, completely conquered the difficulty which led to previous failure, and I am now able to produce a perfectly durable electric lamp by means of incandescent carbons.

Edison, reading Swan's letter, commented with unjustified skepticism: "There you have it. No sooner does a fellow succeed in making a good thing than some other fellows pop up and tell you they did it years ago." Then, claims Francis Jehl, Edison and Upton spent two days searching for details of Swan's work until they at last discovered an article about him in the *Scientific American* of the previous July. To judge by Jehl, Edison had got rather out of touch.

It seems clear that Swan wrote to *Nature* primarily to stake his scientific claim and he cannot have been happy that the same issue of the journal quoted the London *Times* as saying: "Mr. Edison's light is bright, clear, mellow, regular, free from flickering or pulsations, while the observer gets more satisfaction from it than from gas." Early in 1880 Swan formed The Swan United Electric Light Company Ltd., thereby opening up in competition with Edison. Something more than scientific priority was now involved: the infringement of patents.

Both men took a commonsense view of the situation and in October 1883, their companies finally merged to form the Edison and Swan United Electric Company Ltd. However, one problem was solved only to create another. The new company was relying both on Edison's patent of 1879 and on Swan's of 1880. But Swan had been experimenting successfully in 1879 before Edison's patent was lodged and if these experiments involved, as Swan had always claimed, the use of a carbon filament, then they would constitute what was technically known as "prior use." Edison's patent would thereby be nullified and the gates opened to the eager companies anxious to capitalize on what was quickly becoming an electricity rush. Thus there arose an ironic situation in which only the downgrading of Swan's earlier work would save the prospects of the joint company he had formed with Edison.

Luckily for both men, there was one legally essential difference between their end products. Edison's carbon filament was only one sixty-fourth of an inch in diameter. Swan's was larger, and the court was able to rule that while Edison had used a "filament," Swan's experiments had not done so. Edison's patent was upheld and potential infringers held off.

The importance of the filament's thickness became evident when it was compared with that of the carbon "burners" as they were still called in the lamps produced by Sawyer and Albon Man, two other inventors with whom Edison was to be involved in lengthy litigation. The Sawyer and Man filaments were one thirty-second of an inch across. But Edison's, being half this breadth, quadrupled the burner's resistance and halved its radiating surface. This eightfold increase of the ratio of resistance to radiating surface had an extraordinary effect in practical terms: it reduced by seven-eighths the amount of copper or other metal used to distribute the electricity to the lamps.

The negotiations with Swan, like the rest of the patent litigation which was to follow Edison's success, lay in the future as, during the first weeks of 1880, he immersed himself in a huge sea of research and development work. Early in January he applied for patents covering electric lamps, apparatus for producing high vacuums and a particular kind of lamp and holder. On 28 January he applied for an omnibus patent covering his "system of electrical distribution." Before the year was out no less than another fifty-six patents had been applied for, two-thirds of them dealing with lamps, dynamos, auxiliary equipment, or variations of the distribution system. They included patents for an electric railway which was to engage his attentions while he was developing the lighting system and—typical of his diverse interests and his unwillingness to let a potentially good idea get away—a patent for preserving fruit.

The light bulb itself was the center of interest and Edison was frank in admitting that luck as well as routine trial and error had to be exploited. Thus some while earlier he had told a reporter from the *New York World*:

> I don't know when I'm going to stop making improvements on the electric light. I've just got another one that I've found by accident. I was experimenting with one of my burners when I dropped a screwdriver on it. Instantly the light was almost doubled, and continued to burn with increased power. I examined the burner and found that it had been knocked out of shape. I restored it to its original form, and the light was decreased. Now I make all my burners in the form accidentally given to that one by the screwdriver.

There were many modifications and the appearance of the electric light bulb, as it was soon known, began to change. At the end of 1879 it had been globular with an elongated neck, the filaments extending to the interior. The platinum leading-in wires were sealed to the summit of the interior, the tips on the globe were pointed and hollow, and platinum clamps were used. Within the next few months, the size of the globe was enlarged in the belief that this would increase the incandescence. A kind of white German glass was next utilized at the junction of the wires with the glass in order to make a better seal. This was discontinued after a while and the shape of the bulb was again modified.

Throughout 1880 a multitude of refinements and improvements continued to be made. Edison had appreciated that when a replacement was needed a bulb would have to be changed simply, without the customer having to call in a mechanic, and many attempts were made to solve this particular problem. Wooden sockets were for a while used at the base of the bulb and this in turn gave way to plaster of Paris. Eventually, Edison invented the screw cap, adapted from the screw stoppers of kerosene cans which were almost certainly his early prototypes.

Meanwhile there were continuing efforts to lengthen the life of the bulb and within little more than a year one of his specimens had produced 16 candlepower for 1,589 hours. Although longer life meant that fewer bulbs were needed at any one installation, the spread of electricity so increased demand that in 1882 some 100,000 were made. A decade later the annual production figure was running at 4,000,000; by 1903 it had reached 45,000,000.

Such staggering success could hardly have been forecast, even by Edison himself, as in the bustling months of 1880 research on material for the filament was relentlessly continued. He was now convinced that carbon in one form or another must give the best results and he redoubled his experiments. Cardboard and various kinds of paper were carbonized; so was fish line, plaits of different kinds of thread which had been soaked in boiling tar, coconut hair, cork and many kinds of wood. The answer finally came through his perpetually wide-awake observation. On one of the tables in his home there happened to be a palm-leaf fan, one of those objects with which the ladies of the day kept themselves cool in hot weather.

> I was then investigating everything with a microscope [Edison said later] so I picked it up and found that it had a rim on the outside of bamboo, a very long strip cut from the outer edge. We soon had that cut up into blanks and carbonized. On putting these filaments into the lamps we were gratified to see that the lamps were several times better than any we had succeeded in making before. I soon ascertained why and started a man off for Japan on a bamboo hunt.

Men were in fact sent to many parts of the world in the search for better filament material. Their travels, news of which came readily from Menlo Park, caught the public imagination and the *New York Evening Sun* hardly exaggerated the interest when it wrote of one bamboo searcher:

> No hero of mythology or fable ever dared such dragons to rescue some captive goddess as did this dauntless champion of civilization. Theseus, or Siegfried, or any knight of the fairy books might envy the victories of Edison's irresistible lieutenant.

The quest for even better materials might go on: Edison however had already discovered a good Japanese source of bamboo and arranged that regular shipments should be sent to America. This sufficed until, some years later, bamboo began to be abandoned in favour of cellulose, produced by special ejection techniques.

The story during 1880 was by no means one of unbroken success. Early in the year the essential vacuum pump sprang a leak and for a while the production of bulbs virtually ceased. The explanation was simple and the hold-up had nothing to do with the foundations on which success had been based. But Menlo Park was now being given the round-the-clock attention by the Press later to be reserved for Presidential candidates and major affairs of State. No sign of potential success or failure was missed, and the suspicion that the critics might have been right after all sent the price of the Edison Company's stock tumbling. In the summer of 1879 the $100 Edison Electric Light Company shares had sunk to $20. This had been followed, in the aftermath of the post-Christmas Menlo Park demonstration, by a huge rise. The fashionable senator Roscoe Conkling—recovered from the embarrassment of Edison's phonograph demonstration a few years earlier—let it be known that he was buying shares at $3,000. August Belmont, the banker, said he was paying $3,500 and would buy any number offered at that price. Drexel, Morgan & Company, who already owned a substantial holding, were known to be in the market, while smaller shareholders were hanging on in the hope that the price would reach $5,000. Now, with the news of difficulties holding up production, the price slipped back to $1,500.

Nature, then as now one of the world's most authoritative scientific journals, was sceptical and in a two-page article on 12 February—some six weeks after the Menlo Park demonstration—continued to maintain that the Edison system was impractical.

Accurately reflecting conventional scientific opinion of Edison and his activities, it said:

> We need not refer to the enthusiastic inconsistencies in the *Times* correspondent's accounts. Upon Edison's own data, electricity, instead of costing one-fortieth of the price of gas, costs at least seven-eighths as much, or about thirty-five times as dear as the *Times* correspondent declares. As to the cost of the lamp itself, with its carefully incinerated horseshoe of paper, its glass globe exhausted to one-millionth of an atmosphere, and its platinum connecting wires, we confess we do not know where the work could be done for anything like the cost of a shilling. "The current can be transmitted on wire as small as No. 36 (says the *Times* reporter, who, probably being unaware that the resistance of a yard of such wire is at least half an ohm, avoids saying what length of such wire may be used). With a generating machine in a central station, perhaps a half mile away, the introduction of 400 ohms' resistance would be serious—to the light.

After alleging that Edison was "thirty-five years behind the time in his new invention"—an assertion based on a patent for an incandescent lamp by the Englishman Edward King in 1845—the article revealed the source of the strong bias that was maintained against Edison for much of his life. Inventors, it said, sometimes found it impossible to credit the good faith of their rivals.

> Here the scientific man and the inventor part company, since the habits of accurate thinking and the necessary candour of the scientific method preclude the truly scientific man from ignoring, even for the sake of scientific discovery, that which is already a part of scientific truth. We are doing no injustice to Mr. Edison's splendid genius when we say that it is to the character of the inventor, not to that of the scientific thinker, that he aspires.

While Edison was being lambasted by *Nature*, founded and edited by his friend Norman Lockyer, he was helping to finance its American equivalent, *Science*. Lockyer himself may well have fired Edison with the idea of starting such a journal. There was certainly a need for it. But it is difficult not to believe that Edison regarded it as a useful organ for propaganda. Certainly the attitude of *Science* during its first fifteen months, the period when Edison was underwriting its costs, was strongly in his favor. Thus the second issue contained an outspoken editorial comment on reports from England which suggested that Edison was giving up his work on the incandescent bulb. On the contrary, *Science* said, he would be lighting up some parts of Menlo Park, apparently on a permanent basis, within a few weeks.

> This will be Edison's answer to all the meretricious arguments and scientific hair-splitting which has been of late, with little generosity, carefully disseminated to his disadvantage. Taking the view that it is a waste of time to argue theoretically on that which can be demonstrated practically, Edison, through all this wrangle, has been silent but not idle; while others *talked* he has *worked*, and in a few short weeks all will be ready, when those who are competent can see and judge for themselves.

True enough; but less significant to readers if Edison's underwriting of the journal had been known.

Science failed to prosper and in October 1881 Edison withdrew support. He felt, in the words of his secretary, Samuel Insull, "that it is a very poor investment, see[s] no prospect of a return and therefore thinks it advisable to close the matter up before incurring a much greater loss."

Meanwhile he had taken headquarters on Fifth Avenue and from here, as well as from Menlo Park, directed the advance of the electric light along two separate lines. There was first the perfection of equipment needed for the installation of electricity in separate homes, houses or business premises. Secondly, and far more important in his eyes, there was development of the central generating station on which he had set his heart. Most of the work was carried out at Menlo Park where, by the autumn of 1880, a line of street lamps stretched for nearly a mile. In Edison's own house the hall, the parlor and the dining room were lit by sixteen lamps. "In the parlor," noted one visitor, "the effect was particularly fine. Two handsome chandeliers, of four branches each, held eight lamps whose refulgence was thrown downwards by conical porcelain shades that had no opening at the top, none being needed, and all the light being utilized."

By this time competitors were entering the market. All had benefited from Edison's demonstration that the incandescent lamp was practical and some had benefited from his patents although in the confused situation then existing it is difficult to estimate how much of this was deliberate infringement and how much the almost inevitable outcome of men working simultaneously along parallel lines of research. Edison was usually outspoken in his convictions. As an example he maintained that Hiram Maxim's lamp in which the filament was heated in an attenuated atmosphere of hydrocarbon vapor, was "a clean steal" from his own lamp. "I do not worry about them, or a hundred like them," he went on. "I always expected them and there will be more like them. The lamp is to a system of electric lighting what a gas burner is to a gasworks." To the protest that there were reputable names in Maxim's company, he replied:

> All the better for the other fellows. What does a capitalist know when a man shows him a patent with a full-blown picture of an electric lamp on it. The claim may be for a stopcock or a piece of wax, but the whole lamp is in the drawing and the capitalist thinks the fellow has a patent for everything in the picture.

And to the question of how Maxim "got" his lamp his reply was as simple.

> Well, my lamp is no secret. Mr. Maxim came here himself and spent an entire day, from morning until late at night, looking over the whole place. Then he got hold of one of my glassblowers, and there's the whole thing in a nutshell.

While Edison thus seemed confident enough to the Press, the pessimists did appear to have an arguable case when they questioned the possibility of lighting up even a sizable part of a large city from a single generating station.

At Menlo Park the lighting was over a comparatively small area; and, since it had been organized primarily as a demonstration, little attention had been paid to the cost, a major item of which was the copper wiring used to distribute the electricity from generator to bulbs. In order that each light could be switched on or off independently of the rest, they had been wired not in series but in parallel. In theory there were two ways of ensuring that the lights farthest from the generator should burn as brightly as those near to it. The first was to use lamps of different voltages but this was automatically ruled out by the need, in any system such as Edison envisaged, to use lamps which were interchangeable. The alternative was to vary the diameter of the conducting wire. At the far end of the system it could be thin enough to serve merely the lamps connected there. Nearer to the generator it would have to be proportionately larger while an even thicker wire would have to feed the lamps nearest to it. The varying thickness of the conducting wire could be utilized so that 110-volt lamps—or lamps of any other voltage—burned with equal brilliancy wherever they were in the system. But if so, as the critics pointed out, the cost of the copper would be so enormous as to make the system wildly uneconomic. However, Edison was well aware of the situation and had been working at the problem since the start of research on the incandescent bulb. While the critics were still sounding their warnings he was preparing to patent the Edison feeder system.

This was a solution having the basic simplicity of genius. Instead of a district being served by one main conducting wire, each district was divided into smaller areas; in each of the smaller areas there was laid a separate main serving a limited number of lights; and each main was connected to the central generator by a number of feeder wires. The beauty of the system lay in the fact that the long feeder wires running out from the center of the district could be comparatively thin and only small sections of the mains had to be thicker. The result was surprising. In one typical area of New York for which Edison hoped to provide electricity it was found that the copper required by the feeder system would be less that one-sixth of the amount required by the conventional system.

In the early days a fairly primitive method was used for determining the correct thickness of the wires in any part of the system.

A huge map was prepared, showing the location of the streets, and the position of the houses where current was to be supplied [a former member of Edison's staff later explained]. On this map a spool of German silver wire was located wherever a house was to be supplied with lights. Each spool had a resistance proportional to the resistance of the lamps in the house. Wires corresponding to the feeders to be actually used were stretched along the streets, and the German silver spools were connected to these wires. Current was obtained from a small Daniel battery, and distributed to the different spools through the wires. A professor then sat in front of the map and measured with a galvanometer the drop along each of the wires. From his measurements, the proper wires for running along the streets of the city could be determined.

Eventually F. J. Sprague, later to become the pioneer of a successful electric rail system, joined the Menlo Park staff. He showed how the best size for the feeders could be calculated without any experimental work with models.

The feeder system, however, was merely the first of two inventions used by Edison drastically to reduce the amount of copper, thereby making it possible for electricity to compete economically with gas. The second, developed two years later, was the three-wire system, an ingenious arrangement in which the potential difference between the normal main conductors was doubled and a central neutral conductor was added to the circuit. The cross-section of the conventional mains was thereby reduced by 75 percent and while the addition of the third wire increased the amount of copper required by $12\frac{1}{2}$ per cent, this still meant a total saving of no less than $62\frac{1}{2}$ percent.

It was a simple solution. So simple that years later when Sir William Thomson was asked how everyone else had missed it, he replied: "The only answer I can think of is that no one else is Edison." This was a little hard both on Charles F. Brush, who had indicated the system which Edison developed in a patent lodged nine months before Edison's, and on John Hopkinson, earlier employed by Edison's telephone company and eventually consultant to his English lighting company. Hopkinson had worked out a similar scheme quite independently of Edison and in England the three-wire system was patented jointly in both their names.

Reducing the amount of copper to be used in the distribution system was only one task. Edison realized that if his plans were successful it would be impractical to string electric wires as telegraph wires had been strung, on long poles along the city streets. On the other hand if they were to be buried underground then some foolproof method was necessary to prevent the electricity from seeping out into the surrounding earth. A fraud-proof system of measuring the amount of electricity used had to be devised, and so did a means of ensuring that an accidentally large flow of current did not burn out scores of lights. Yet meters, fuses, switches, underground distribution wires and the connecting boxes to go with them, were only a few items of equipment demanded by the Edison system, taken for granted today but virtually unknown in 1880.

It was at this stage that the virtues of the Menlo Park organization became apparent. At the end of 1878, when Edison had achieved his first, if transitory, success with the incandescent bulb, he had said: "Now that I have a machine to make the electricity, I can experiment as much as I please. I think this is where I can beat the other inventors, as I have so many facilities here for trying experiments." If this was true of the bulb, it was even more true of generation and distribution. While most pioneers of the electrical industry had at their disposal only the resources of their own specialist laboratories, Edison had at hand the prototype of the modern industrial research department. At Menlo Park the process of "invention" had already begun to merge into "development"; here it was no longer a question of one man seeing with a flash of insight how a problem could be solved but of a team worrying away at it until a solution was found by the process of trial and error. It was a radically new method of tackling technological problems made possible by the range of facilities available. If Edison wanted to test material for a specific purpose, the means were probably to hand. If he required a new piece of equipment he could order and supervise its design and its manufacture. Upton could supply theoretical calculations, Boehm in the glassblowing section could produce bulbs, or other glass vessels, at short notice. For most of what would today be called "research and development," Menlo Park was a self-contained unit.

The self-sufficiency does much to explain the speed with which Edison and his colleagues were able to create what was later described as "a central electricity lighting plant with the multitude of practical devices that are necessary." Of first importance was the dynamo. Ever since 1831 when Faraday had shown that it was possible to convert mechanical energy into electric current, men had designed equipment to do this. Yet even the most efficient dynamos yielded as electric current only about 40 percent of the energy fed into them. Edison was not the man to lose 60 percent of input before distributing his electricity, and design of a more efficient electricity producer was among his first tasks.

The problem was tackled in the typical Edison manner. Armature cores were constructed of cast iron, forged iron and sheet iron, and then tested to discover which was most efficient. The armatures themselves were built in a variety of shapes and each then tested. Various forms of armature winding were experimented with and so were differing ways of dividing up the armature and the commutator. The outcome was construction of "Long-Waisted Mary Ann," weighing 1,100 pounds and the largest dynamo then in existence. After the first tests, Edison proudly announced that it showed an efficiency of 90 percent, a figure met with incredulity by many experts. In fact efficiency turned out to be 82 percent, an astonishing enough figure since it meant that in one bound Edison had doubled the efficiency of what was to become the main method of producing electricity.

Other problems were solved by the same process of methodical experiment. Scores of different materials and methods were tested until the one best suited to the purpose was discovered. His attack on the seepage of electricity which he foresaw when his wires were buried underground was typical. First, he ordered one of his staff to read up all the available literature on insulation. Next, each of the more likely materials had to be tested. The outcome was a long line of kettles in one of the Menlo Park laboratories, each boiling up one of the compounds to be tested.

The results of this stew were used to impregnate cloth strips, which were wound spirally up on No. 10 B.W.G. wires one hundred feet in length [Francis Jehl has written]. Each experimental cable was coiled into a barrel of salt water and tested continually for leaks. Of course there were many failures, the partial successes pointing the way for better trials.

Eventually it was decided that asphalt, boiled in oxidized linseed oil with paraffin and a little beeswax, gave the best insulation and this was then used to cover the bare cables that had been laid beside trenches dug in the village streets.

> Barrels of linseed oil, bales of cheap muslin and several tons of the asphaltum were hauled in [says Jehl], two fifty-gallon iron kettles were mounted on bricks, and the mixing operation was soon progressing in a big way. Through the pot in which this compound was boiled we ran strips of muslin about two and a half inches wide. These strips were bound into balls and wrapped upon cables. After the man who served these tapes upon the cables had progressed about six feet, he was followed by another man serving another tape in the opposite direction, and he in turn by a third man serving a third tape wound upon the cable in the direction of the first winding. After the cables were all covered with this compound and buried, the resistance to the earth was found to be sufficiently high for our purpose.

Another problem in the early days was raised by fear of a consumer overloading the system. A safety device was concocted, called by *Nature* "a most ingenious contrivance whereby if any consumer draw too largely on the supply the armature of an electromagnet in the circuit is attracted and 'cuts out' the transgressing consumer, actually fusing up the only remaining metallic connection." This drastic remedy was eventually not needed but it nevertheless remained necessary to discover how much electricity each subscriber was using.

Edison had abandoned his early scheme of merely charging a lump sum for every lamp used. Each of the various metering ideas that was tested had its own problems and as late as the autumn of 1880 he was still searching around for a solution. Although his staff worked very largely as a series of teams, there were at times difficulties about putting up original suggestions to Edison himself; the meter problem was an example. This was discovered by Edward G. Acheson who had been casually hired only a short time before.

> What appeared to be a happy thought occurred to me for the method and design of a meter. I made a drawing of my proposed instrument, and the next time Edison came into the room I showed it to him. He seated himself on a high stool at the drawing table, put his arms on the board and his head, face down, on them, and seemed lost for some time in thought. After some minutes he raised his head and addressing me said, "I do not pay you to make suggestions to me. How do you know but that I already had that idea, and now if I use it you will think I took it from you."

While he was working for Edison, Acheson explained, he felt that anything he thought up belonged to his employer, a reply that appears to have been satisfactory. The meter was built and tested. Like all previous devices, it was a failure and Edison went on to develop his own.

The Wizard of
Menlo Park

"A bloated Eastern Manufacturer"

Below: *Sally Jordan's boarding house at Menlo Park, one of the first houses to be lit by electricity. Mrs. Jordan (in black) stands on the porch at left with her daughter Ida beside her. The boy standing near the lamp post (center) is Thomas A. Edison, Jr., and the rest of the group are Edison employees*

The first photograph taken by electric light. The date is 1883 and the subject Charles Batchelor

Right: *A replica of Edison's first successful incandescent lamp*

The Edison Machine Works at Goerck Street, New York, 1881

Below: *Laying electric cables to supply New Yo buildings with light from Edison's Pearl Street generating station*

aying the Electrical Tubes

MACHINE SHOP.

Above: *Edison's first generator, "Long-Waisted Mary Ann"*

Left: *Edison (left) outside the offices of the Edison Electric Light Company, 65 Fifth Avenue, New York, in 1881. Charles Batchelor is standing (center), with (on right) Major S. B. Eaton, President of the company*

*The Edison Exhibit
at the Crystal Palace Exhibition,
London, in 1882*

THE EDISON EXHIBIT

118

SON DYNAMO MACHINES.

Below: *Edison's "Jumbo" dynamo. Site unknown, but probably the Holborn Viaduct station, London*

Edison's electric railroad, built at Menlo Park in 1880

The Palace of Electricity,
Champ de Mars, at the
International Exhibition,
Paris, 1900

Right: *Two pages from
Edison's notebook for
February 1880. Edison
continued to improve the
light bulb throughout
that year*

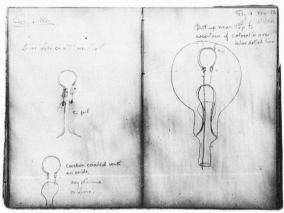

*Edison with two of the lamps
produced to illustrate "the Edison effect,"
later to become the basis of the
electronics industry*

"Glenmont," Edison's home at Orange, New Jersey, for the later years of his life

Right: *Edison with his three children by his second wife, 4 July 1900. "Our Fourth of July was father's special day," his daughter says. "Here he is trying to get some reaction from his young son, Theodore, to a fire cracker which is about to explode. Theodore, however, calmly awaits results"*

Left: *Mina Miller Edison, Edison's second wife, photographed about 1894*

Right: *Edison with his son, Charles, after a day's fishing in Florida, about 1900*

The "Napoleon" photograph of Edison taken at his labor[atory], Orange, New Jersey in June 1888, after he had worked continuously for some days on developing the phonograph

Below: Edison and his workers on the same day. (Standi[ng], left to right) W. K. L. Dickson; Charles Batchelor; A. T. E. Wangemann; John Ott; Charles Brown. (Sitting, [left] to right) Fred P. Ott; Edison; Colonel Gouraud.

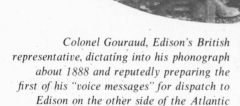

Colonel Gouraud, Edison's British representative, dictating into his phonograph about 1888 and reputedly preparing the first of his "voice messages" for dispatch to Edison on the other side of the Atlantic

Colonel Gouraud's secretary, "the first British audio-typist." Although finding short-hand difficult to master, she earned a living as a "lady typewriter" with the help of the phonograph as a dictating machine

A recording session about 1888–89 at Little Menlo, the Upper Norwood, home of Colonel Gouraud; believed to have been the first musical recording session held in Britain

Below: A distinguished gathering in Gouraud's London apartment listening in the early 1890s to a recording by the late Cardinal Newman. Seated near the instrument is Lady Herbert of Lee and behind her is Lady Jeune (Lady St. Helier). On the extreme right is the Duchess of Teck, mother of Queen Mary. Others present include H.M. Stanley, the explorer; Sir Ellis Ashmead Bartlett, the British politician born in Brooklyn; Lord Aberdeen, Governor General of Canada; Sir Charles Russell; and Cardinal Vaughan

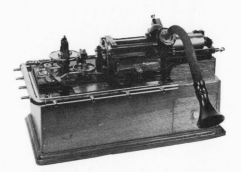

Opposite: *Edison listening to a new disk phonograph about 1909*

Above: *The Edison "Class M" Phonograph, about 1893. Powered by a battery-driven electric motor. The speaking tube is for recording on to a cylinder and the "gallery" with nozzles is for hearing tubes, which enabled a number of people to listen at the same time*

*he Edison Idelia, introduced about 1910.
his de luxe phonograph played two-
d four-minute cylinders, and the
producer had twin styli with
turn-over device like modern
cord players. This particular
odel was also fitted with an
utomatic repeating device
that, once set, the same record
ould play repeatedly without
tention*

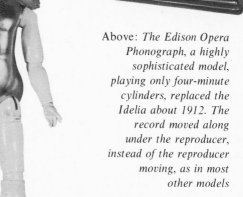

*ight: Edison's
eaking doll,
veloped in the 1880s,
ntained a miniature
onographic cylinder*

Above: *The Edison Opera Phonograph, a highly sophisticated model, playing only four-minute cylinders, replaced the Idelia about 1912. The record moved along under the reproducer, instead of the reproducer moving, as in most other models*

127

A recording session being held on
30 March 1916 at the Edison Studio, 79 Fifth Avenue, New York.
Jacques Urlus singing, Sodero conducting

128

His final solution invoked electroplating, in which a known weight of metal is transferred in a given time from the positive to the negative plate by each ampere of current. A small fraction of the current which passed into each customer's premises was diverted through a pair of cells, each consisting of two zinc plates immersed in a solution of zinc sulphate connected in series and thus acting as a check on each other, and the amount of current used was directly proportional to the amount of zinc removed from the positive plates and deposited on the negative. Thus to find how much electricity had been used it was only necessary to weigh the plates.

Edison's relations with the men who worked for him was characterized, as much in Menlo Park as in Newark, by a gruff friendliness. Acheson, for instance, had walked out after being refused higher pay and had then returned to the laboratory. Edison laughed off the walk-out, turned to Acheson and said:

> There in the end of the room is a hydraulic press; have it put in order, and make for me a small graphite loop like this (making a sketch like a horseshoe). I want the loop one inch outside diameter, the filament to be twenty-five thousandths of an inch wide and two-thousandths of an inch thick. I will have steel plates made for you to press sheets between, and a die made for punching out the filaments. When you make one capable of mounting in a lamp, I will give you a prize of one hundred dollars.

Acheson made the filament, received his hundred dollars, and was then offered by Edison a contract to make 30,000 more. "I engaged a man and a boy to help me," he wrote, "and became so expert at making them that I was earning twelve dollars a day by the time 16,000 of them had been turned out."

However, while the filaments produced a brilliant light, they did not last very long. Acheson felt it was a waste of money to go on making them, told Edison, and was rewarded by being put through a course of lamp manufacture and then sent to Europe as an expert.

Once the practicability of the incandescent lamp had been demonstrated beyond doubt Edison was inundated with requests to install individual lighting plants. Soon, however, a rift opened up between him and his backers. The bankers, concentrating on what they saw as secure short-term successes requiring comparatively small capital, supported the idea of isolated lighting plants which would serve a factory or a block of offices. Edison, seeing ahead to the future, wanted to concentrate on the larger central station that would serve an entire area of a large city. The two ideas were, in fact, pressed on side by side.

However, before he could forge ahead with his central station idea, Edison found it necessary to start a separate clutch of companies—the Edison Lamp Company, the Edison Machine Works, the Edison Shafting Company, the

Thomas A. Edison Construction Department and the Edison Electric Illuminating Company of New York which eventually built and operated the station. New premises had to be acquired, including a site on Goerck Street where the dynamos were to be built. Although some money was raised for these in New York's banking world, Edison was forced to support the ventures himself by selling many of his original 2,500 shares in the original lighting company of which he thus lost control. Wall Street backed the incandescent bulb but it was Edison who backed the ideas without which it might never have been brought into the mass of private homes.

As long as work on the central station idea was not hampered, he was not averse to installing individual plants and in fact the Edison Isolated Lighting Company was set up to license such installations in cities throughout America. He himself installed a number, particularly in New York, where they were useful either for gaining experience or for publicity. Thus when Hinds and Ketcham, a New York firm of lithographers, came to consult him, Edison was cooperative. The firm had difficulty in color printing by artificial light and this limited their work to the hours of daylight. Edison, anxious to show what the new lamps would do, installed an isolated plant without delay.

There was also James Gordon Bennett of the *New York Herald* who a decade earlier had sent out Henry Stanley to find Livingstone. Edison was anxious to light a newspaper office and had installed a printer's composing case at his headquarters on Fifth Avenue so that any visiting proprietor could be shown the advantages of typesetting by electric light. Gordon Bennett was the first to come, and at once ordered his general manager to switch to electricity.

It was also with Gordon Bennett's introduction that George Washington De Long, the young naval officer and Arctic explorer, had visited Edison at Menlo Park in 1879. After the Swedish scientist, Baron Nordenskjöld, had found the long-sought northeast passage to Asia by sailing from Norway to the northern entrance of the Bering Strait, De Long, backed by Bennett, planned to sail south to north through the Strait. Edison supplied the vessel, the *Jeannette*, with a small dynamo and incandescent lamps as well as hand-operated telephone sets, linked by long cables, which would enable parties landed on the ice floes to keep contact with each other.

The dynamo, Edison's first, does not appear to have been of much use judging by the two references to it in De Long's journal. After the first efforts with it had failed, De Long noted that it appeared to be "irretrievably worthless." Further attempts were no more successful. A few days later, he wrote:

> The electric machine after having received [the engineer's] attention, had been
> in hand for some days unreeling and reinsulating, and reeling again the wires, and

was ready for another trial. Steam was accordingly raised in the Baxter boiler, and the generator connected; but though seventy pounds steam was applied, not a spark even could be obtained, nor a deflection in the galvanometer needle. The only effect was to fill the deckhouse with the fearful smoke of burning blubber, and to make it dripping wet from condensing steam and the shower of rain falling from the roof.

The electric equipment was finally abandoned and they had no better luck with the communication system. The wires, De Long reported, were "broken in several places this morning from their own weight, increased by a slight amount of frost. We have tried laying them in the snow, but it has rotted them through and through."

De Long sailed through the Bering Strait without trouble but was caught in the ice fields further north, and of the crew who tried to make the long overland trek back to civilization less than half survived.

The first, and unfortunate, attempt to use the incandescent lamp at sea was followed by success. Early in 1880 Henry Villard gave Edison his first order for commercial installation of an electric plant. It was to be on the S.S. *Columbia*, a steamship of the Oregon Railway and Navigation Company of which Villard was by now President, and despite warnings that the vessel would be set ablaze the *Columbia* left New York in May 1880, sailed round Cape Horn and reached San Francisco without incident, all lights working perfectly throughout the voyage. Almost two years later the company's chief engineer sent a eulogistic report, extolling the safety, cleanliness and convenience of the equipment and adding, "the engines being connected to the main condenser when under way, the actual expense felt consists only in the extra pint of oil used in lubricating engines, dynamos, etc."

Edison also installed incandescent systems in a small number of private houses, notably that of Pierpont Morgan at 219 Madison Avenue, the first home in New York to be lit throughout with the new bulbs. From the start there were complications and the Morgan family's experiences give an idea of the teething troubles with which Edison had to deal.

A cellar was dug underneath the stable which stood on Thirty-sixth Street in the rear of the house, and there the little steam engine and boiler for operating the generator were set up [says Morgan's biographer]. A brick passage was built just below the surface of the yard, and through this the wires were carried. The gas fixtures in the house were wired, so that there was one electric light bulb substituted for a burner in each fixture. Of course there were frequent short circuits and many breakdowns on the part of the generating plant. Even at the best, it was a source of a good deal of trouble to the family and neighbors. The generator had to be run by an expert engineer who came on duty at 3:00 P.M. and got up steam, so that at any time after four o'clock on a winter's afternoon the lights could be turned on. This man went off duty at 11:00 P.M. It was natural that the family should often forget to watch the clock, and while visitors were still in the house, or possibly a game of cards was going on, the lights would die down and go out. If they wanted to give a party, a special arrangement had to be made to keep the engineer on duty after hours.

Neighbors complained of the noise made by the dynamo, of the vibrations, and of the smoke from the generator. Complaints ceased only after the machinery had been remounted on rubber pads and sandbags had been piled round the walls of the cellar.

One disadvantage of installing equipment at this stage of development was that it so quickly became out of date. Less than two years after the installation at No. 219 Morgan asked Edison for advice and was visited by Edward Johnson, now Edison's righthand man. Johnson sized up the situation and asked Morgan if he could be frank. Told to speak up, he said: "If it were my own, I would throw the whole damn thing into the street." Mrs. Morgan, it seems, had said substantially the same thing.

Whatever troubles the Morgan installation might raise, the fact that New York's leading banker was lighting his home by electricity was of enormous publicity value, a factor which Edison rightly enough kept high on his list of priorities as he struggled against the entrenched forces of the gas companies. For this reason, if for no other, he welcomed the visit to Menlo Park of Sarah Bernhardt at the end of 1880 and ensured that it would have the maximum effect.

Bernhardt was playing in New York and it was two in the morning before she arrived. Edison had ordered that all lights in the area should be put out and all was dark except for the oil lamp on one of the two carriages sent to meet the party. It was a black December night, snow was falling, and with some justification Bernhardt felt that the visit strongly resembled a scene from an operetta.

I cannot tell [she recorded] how long we had been rolling along, for, lulled by the movement of the carriage and buried in my warm furs, I was quietly dozing, when a formidable, "Hip, hip, hurrah!" made us all jump, my travelling companions, the coachman, the horses, and I. As quick as thought the whole country was suddenly illuminated. Under the trees, on the trees, among the bushes, along the garden walks, lights flashed forth triumphantly.

Following this flamboyant Edison touch, Bernhardt was given the full treatment, an apparently dour reception from the man himself which so melted that "within half an hour we were the best of friends," and a conducted tour

132

of the laboratories with lights flashing on and off at the commander's orders. Almost inevitably, she compared him to Napoleon.

> The deafening sound of the machinery, the dazzling rapidity of the changes of light, all that together made my head whirl, and forgetting where I was, I leaned for support on the slight balustrade which separated me from the abyss beneath. I was so unconscious of all danger that before I had recovered from my surprise Edison had helped me into an adjoining room and installed me in an armchair without my realizing how it had all happened. He told me afterwards that I had turned dizzy.

Supper was followed by a chat with the family, and finally departure at four o'clock in the morning. There can be little doubt that, back in Paris, Sarah Bernhardt declaimed about the wonders of Menlo Park.

Chapter Seven

The Universal Lamplighter

Even by the summer of 1880 Edison knew that the incandescent lamp was well along the road to success, whatever problems still lay ahead. He could see not only the shape but the size of things to come and to help in the organization of what he knew would be a vast series of enterprises he pulled back Edward Johnson from London. The Edison and the Bell telephone interests had just been satisfactorily amalgamated and it was evident that Johnson would now be of more use at Menlo Park.

Johnson's return to America was to be significant for Edison's fortunes in one way that neither man can have suspected at the time. Slightly more than a year earlier Colonel Gouraud, who had been so ably fighting Edison's telephone battles against Bell's Colonel Reynolds, had engaged as secretary Samuel Insull, a young Englishman. Insull had two years earlier read "A Night With Edison," the article in *Scribner's Monthly*, and it had turned him, overnight, into an Edison worshiper. He had quite accidentally stumbled into the job with Gouraud without knowing of the Colonel's connection, had been overwhelmed on discovering it, and had thereupon concentrated all his efforts on joining the great man himself.

The chance came soon after the return to America of Johnson, for whom Insull had occasionally worked part-time. Edison's current secretary resigned in January 1881, and Insull was brought over to take his place. The *City of Chester* on which he sailed from Liverpool arrived in New York at dusk, and the young man was taken, without delay, to start work for Edison at 65 Fifth Avenue.

Insull's biographer gives a revealing account of the meeting.

Insull looked at Edison, and Edison looked at Insull, and both were disappointed; each said to himself, "My God! He's so young!" Insull was twenty-one and looked sixteen; he was skinny, popeyed, fuzzy cheeked; his dress was impractically impeccable and his manner was impractically formal. His cockney accent was so thick that Edison had difficulty understanding him. Mouth agape, he stared back at his hero. Edison had just turned thirty-four. He wore a seedy black Prince Albert coat and waistcoat, black trousers that looked as if they had been slept in, a large sombrero, a rough brown overcoat, and a white silk handkerchief around his neck, tied in a careless knot and falling over the front of a dirty white shirt. His hair was long and shaggy; he was beardless but ill-shaven. His manner was as casual as Insull's was formal. His Middle-Western accent was so thick that Insull had difficulty in understanding him. Overall, the only saving feature was "the wonderful intelligence and magnetism of his expression, and the extreme brightness of his eyes."

Yet here, facing one another for the first time, was the greatest inventor of the century and the man who was to bring business order out of inventive chaos. Long before the meeting in February 1881, Edison had made more than one considerable fortune. But during the next twelve years, as the success of his electrical enterprises demanded increasingly complex financial operations, he could well have foundered. The somewhat happy-go-lucky conditions under which the robber barons had made and lost fortunes were slowly changing; a more sophisticated form of expertise was required and by a stroke of luck Edison was now able to hire someone who had it in plenty. In return, he was to imbue the young Insull with some of his own individual business philosophy. Thus half a century later Insull—in the box on a multitude of fraud charges of which he was acquitted—remembered how Edison had told him: "Whatever you do, Sammy, make a brilliant success of it or a brilliant failure. Just do something. Make it go."

Insull arrived shortly after Edison had taken a decisive move forward in his progress toward a central generating station. Only a company acting under the city of New York's gas statutes could operate an enterprise of this kind and on 17 December 1880, the Edison Electric Illuminating Company of New York was formed for this purpose. There were now five companies preparing to introduce various forms of electric light into New York, those of Brush, Maxim, Jablochkoff, Sawyer and Gramme, and Edison. Success would rest not only on the intrinsic merits of the bulbs but on the efficiency of their back-up systems, as Edison was explaining to Insull within an hour or so of their first meeting.

> Right after dinner we sat down and [he] explained that it was necessary for him to start three or four manufacturing establishments to produce the dynamos and lamps and underground conductors which would be required for the installation of the first district of the electric light system in New York. He produced a wallet from his pocket, and told me that he had $78,000 to his credit at Drexel, Morgan & Company, and he asked me where he could get the balance.

Insull had taken care to acquire a thorough knowledge of Edison's financial affairs and readily supplied the answers. With cash soon available, and with companies for making the ancillary equipment being organized, Edison chose the area of New York he intended to serve first. He wanted it to include private houses as well as businesses, preferably some banking houses and also some factories which might be willing to use electricity for power as well as for lighting.

> It occurred to me one day [he later said] that before I went too far with my plans I had better find out what real estate was worth. In my original plan I had 200 by 200 feet. I thought that by going down on a slum street near the waterfront I would get some pretty cheap property. So I picked out the worst dilapidated street there was, and found I could only get two buildings each 25 feet front, one 100 feet deep and the other 85 feet deep. I thought about $10,000 each would cover it: but when I got the price I found that they wanted $75,000 for one and $80,000 for the other. Then I was compelled to change my plans and go upward in the air where real estate was cheaper.

The site was 255–57 Pearl Street. The area that Edison decided to serve from it was bounded by Wall Street, Nassau, Spruce and Ferry Streets, Peck Slip, and the East River. First he carried out a detailed survey of the district, a house-to-house investigation at the end of which he knew how many gas jets burned in the area, for how long they burned on an average day, and how much the gas was costing each establishment every year. Almost as a sideline, his survey showed how many hoists and similar contrivances there were in the area, and thus what the potential was for the supply of electric power as distinct from electric light.

The distribution network, he had decided, should be underground. This meant that New York's streets had to be dug up—and for a form of lighting whose dangers and potential for disaster were naturally being emphasized by the gas corporations. Good public relations work was called for, and shortly after the Illuminating Company was formed the New York aldermen were invited to Menlo Park. They were given a formidable display of what incandescent lamps could accomplish, as well as a sumptuous dinner provided by Delmonico's. Edison's demonstration, decided the *New York Herald*, was "a decided success, especially of his guests' capacity for champagne." He got his permission to lay pipes in New York's streets—despite the attempted veto of the mayor.

While preparations for the central generating station now went ahead, Edison began to win fame abroad for his incandescent system. One of the first important displays was at the first French Electrical Exposition in Paris in 1881. To power the exhibits there, he had built what was then the world's largest dynamo, a twenty-seven-ton giant that could light 1,200 lamps. Business colleagues traveled to Paris and prepared everything for the display but Edison himself remained in New York and supervised the final work on the dynamo. It was just as well. From the start nearly everything went wrong and it needed all his ingenuity and energy to get the equipment to the docks in time. First the voltage generated was found to be too low. Then the crankshaft broke and spun across the width of the machine shop, an indication of the hazards that were still the order of the day. Edison later remembered:

> By working night and day a new crankshaft was put in, and we had only three days left to get it aboard the steamer, and we had also to run a test. So we made arrangements with the Tammany leader, and through him with the police, to clear the street—one of the New York cross-town streets—and line it with policemen, as we proposed to make a quick passage, and didn't know how much time it would take. About four hours before the steamer had to get it, the machine was shut down after the test, and a schedule was made out in advance of what each man had to do. Sixty men were put on top of the dynamo to get it ready, and each man had written orders as to what he was to perform. We got it all taken apart and put on trucks and started off. They drove the horses with a fire bell in front of them to the French pier, the policemen lining the streets. Fifty men were ready to help the stevedores get it on the steamer—and we were one hour ahead of time.

Edison's resulting success at the Paris exhibition was followed by the setting up of a French company to install his electric lighting systems and, shortly afterwards, by other companies in England, Italy, Holland, and Belgium. The advantages of showing his work at major exhibitions was not lost on him. In 1882 he staged at the Crystal Palace, on the outskirts of London, an impressive display whose beauty was recalled thirty-eight years afterward by John Ambrose Fleming, the man whose valve, developed from the "Edison effect," was the basis of the electronics industry. The display included, Fleming remembered, "a large, cone-shaped pendant electrolier, carrying a brilliant equipment of incandescent lamps, and a screen bearing an artificial vine in metal work, the grapes on which were represented by Edison glow lamps in glass shades of floral shape and various colors." The following year it was Munich, while in 1883 Edison also helped light up Moscow for the new Tsar's coronation, using 3,500 lamps on the tower of Ivan the Great and its side galleries.

Early in 1882, the giant dynamo was shipped from Paris to England, immediately named after "Jumbo," the famous elephant in the London Zoo brought to the United States by Phineas Barnum, and used to power the world's first central electricity station on Holborn Viaduct, London. Here, some 3,000 lamps were lit by the station's generator, including the street lamps on Holborn Viaduct itself. The nearby City Temple of the renowned Dr. Joseph Parker became the first church in the world to use the incandescent lamp, while the telegraph room of London's main Post Office a few hundred yards away was lit by 400 Edison lamps. This was something of a personal triumph for Edison since William Preece, one of his earlier and most confident critics, was by now the leading Post Office engineer. Preece not only admitted that he had been wrong on the question of subdividing electricity but was magnanimous, writing in the *Journal of the Society of Arts*:

> Many unkind things have been said of Edison and his promises; perhaps no one has been severer in this direction than myself. It is some gratification for me to be able to announce my belief that he has at last solved the problem that he set himself to solve, and to be able to describe to the Society the way in which he has solved it.

Edison's operations in England were soon to have one important outcome. The electrical adviser to the English company, John Ambrose Fleming, was successful in winning a contract for installing electric light in H.M.S. *Mooltan*, one of the British Admiralty's Indian troopships. The Admiralty was agreeable to the necessary revolutions of the Edison dynamo being obtained by belting from the flywheel of a steam engine, but this was not good enough for the other Admiralty contractors. They insisted that the dynamo should be coupled directly up to a high-speed engine. The Edison company's scientific adviser, John Hopkinson, saw that this method would be impossible with the existing dynamo. However, what he also saw was that by altering the shape of the dynamo's magnet the required voltage could be produced at a lower engine speed. "It is necessary to make a critical study of the [Edison] machines with a view not only to improving them but of placing ourselves in a position to say beforehand how we should modify a machine to meet varying conditions," he wrote to the company. "I am not without hope that we may succeed in largely increasing the output of the machines by alterations of the magnets."

Shortly afterward he built scale models and after tests urged the Edison company to revise the existing pattern. The outcome was the Edison-Hopkinson dynamo, first made in England by the Manchester firm of Mather & Platt. Hopkinson, knowing that the previous dynamo of similar weight could light about 100 lamps when running at 1,170 revolutions, reported on its performance with, as his son wrote, a note of triumph which he rarely permitted himself. The new machine lit 200 lamps.

While improvements were steadily made not only to the dynamo but to almost all the individual pieces of equipment required by the Edison system, work had been continuing in New York on what was to become the Pearl Street power station. By early September 1882, everything was ready and on the afternoon of the 4th Edison gathered with the board members of the Edison Electric

Light Company in the offices of J. P. Morgan at 23 Wall Street. Morgan had been a major backer of the project and his offices were among those to be lit by the 400 lamps already wired into the circuit.

In recognition of the occasion Edison had succumbed to convention, appearing in a new Prince Albert coat, starched white shirt, white cravat, and a white high-crowned derby. He had earlier synchronized his watch with that of the chief electrician at Pearl Street and as zero hour approached he pulled it out. A member of the board wagered: "A hundred dollars they don't go on." "Taken," Edison replied.

The last few seconds ticked by. Then, on the dot of 3:00 P.M., the lights began to glow, becoming brighter as the dynamo in Pearl Street fed current into the circuit. In nearby offices other incandescent lamps blazed out—the first signs of a new era, the culmination of what Edison often called "the greatest adventure of my life." The *New York Times* reported the following day:

> It was not until about seven o'clock, when it began to grow dark, that the electric light really made itself known and showed how bright and steady it is. Then the twenty-seven electric lamps in the editorial rooms and the twenty-five lamps in the counting rooms made those departments as bright as day, but without any unpleasant glare. It was a light that a man could sit down under and write for hours without the consciousness of having any artificial light about him. ... The light was soft, mellow, and grateful to the eye, and it seemed almost like writing by daylight to have a light without a particle of flicker and with scarcely any heat to make the head ache.

There were only a few minor hitches on the first afternoon of the electric light age. At one place a fuse blew out and—almost inevitably—it was Edison himself who insisted on tracking down the fault, climbing down into a manhole where he was discovered by a reporter disarrayed, collarless and his white derby stained with grease.

It was a typical Edison touch—a genuine anxiety to get to grips with any problem that cropped up rather than the flamboyant desire for the limelight that it sometimes looked. Thus there was later the occasion when he visited the opening night of a Boston theater which had just been turned over to electricity. Edison, his colleagues, the Governor of Massachusetts, and all his staff were in full evening dress. During the second act the lights began to dim. The theater's boiler man had detected a fault and while dealing with it had forgotten to keep the boiler stoked. Soon afterward Edison and his secretary were discovered, their tailcoats and silk hats hanging on a peg, shoveling fuel into the boiler.

The day after the opening of Pearl Street, Edison visited his customers to see how things were going. Among them was a Mr. Kolb, who was asked how he liked the new light. "I replied," he said afterward, "that they were all right but that you couldn't light a cigar from them. Mr. Edison said nothing but three days later he came back and presented me with an electric cigar lighter."

Within a few days of Pearl Street starting up Edison achieved another record. This was the first hydroelectric plant—for long claimed to have gone into operation before rather than just after Pearl Street—the forerunner of such giants as the Hoover, Grand Coulee, and Bonneville installations which were to supply millions of watt-hours. Powered by the water of the Fox River, it turned a dynamo at Appleton, Wisconsin, lighting between 200 and 300 lamps in the neighborhood.

At last the critics were beginning to be silenced. However, the production of electricity was still in a primitive state. The unexpected happened frequently; rough-and-ready measures were often called for to cope with emergencies, and there remained, among the public at large, a fear of electricity that amounted to fear of the unknown comparable to, if less justifiable than, contemporary fears about nuclear power. Something unexpected, even by Edison, had taken place during an early test of the dynamos being built at Goerck Street. One engine was put on, Edison recalled.

> Then we started another engine and threw them in parallel. Of all the circuses since Adam was born we had the worst then. One engine would stop and the other would run up to about a thousand revolutions; and then they would seesaw. The trouble was with the governors. When the circus commenced the gang that was standing around ran out precipitately, and I guess some of them kept running for a block or two. I grabbed the throttle of one engine and E. H. Johnson, who was the only one present to keep his wits, caught hold of the other and we shut them off.

Experience now came to the rescue. Edison diagnosed that one set of governors was running the other, gathered in men from the machine shop and quickly had them at work on a shafting device which he expected to cure the trouble. When it failed he took matters into his own hands, went down to the Goerck Street works and improvised his own shafting and tubes that solved the problem.

By the end of 1882 the Pearl Street station was supplying 231 customers and lighting more than 3,400 lamps. By mid-August 1883, when the plant had been operating for almost a year, the Edison Electric Light Company reported that it was serving 431 houses with more than 10,000 lamps.

Speed of making the bulbs increased and the time needed to exhaust them, at first between four and six hours, was soon down to about an hour. At the same time the cost of making the "hot hairpins in bottles" continued to decrease just as their life expectation continued to lengthen.

The cost did not affect the customers since Edison had audaciously priced the first lamps at forty cents each even though they cost $1.40 to make, and did not change the price for years. There was good sense in this, since the cost of manufacture dropped first to seventy cents, then to fifty and finally to twenty-two cents. What did affect customers was the increased average life of the bulbs, about 400 hours when the Pearl Street station was switched on, but quickly increased.

Despite the apparent success of the enterprise there were to be occasions during the next few years when financial disaster looked uncomfortably close. Samuel Insull later recalled how he had gone into Edison's office one night after supper to find out how they were to raise money for the next payday.

> He remarked that it looked a pretty hopeless proposition: he thought that we were at the end of our rope. He said: "Sammy, do you think you can earn a living again as a stenographer?— because if you do I think I can earn my living as a telegraph operator. So that we can be sure of having something to eat anyway."

Whatever the fortunes of the Pearl Street station, isolated plants were soon being ordered in numbers. By the spring of 1883 more than 300 were in operation and it was with mixed feelings that Edison saw these pulling in profits, as the bankers had forecast, on a scale that Pearl Street could not achieve. A music hall, the *Pittsburg Times'* building, cotton mills, flour mills and grain elevators, were only some of the places supplied with individual plant. Numerous steamers were equipped, and a dynamo to light 150 lamps was supplied to Jay Gould's new yacht *Atlanta*. Edison was already becoming what the *New York Times* was to call "the universal lamplighter."

Some of the installations were complex. In Chicago for example Haverly's Theater was equipped with 637 lights, including those on a $500 chandelier hanging in the center of the auditorium. Separate dynamos served the lights in vestibules and entrances, on the stage, and in the auditorium. "Lights in any part of the house can be lowered, raised or entirely extinguished," noted the *Operator & Electrical World*. "In the dressing rooms the lights are independent and can be turned on and out the same as gas. Each burner is warranted for 600 hours, and will last three or four months in a theater unless broken. Of course this light does away with all danger from fire." Some actors had feared that their make-up would look wrong under the new light, but the fear vanished with the introduction of electricity in the dressing rooms as well as on stage. "The expense of putting in this light has been $15,000 to Manager Haverly," it was added, "but he was determined to do away with gas."

Hotels also began to make use of electricity, the first being the isolated Blue Mountain Hotel in the Adirondacks. Some 3,500 feet above the sea, it was in the 1880s forty miles from a railroad, and machinery for the installation was carried to the site on mules. Transport of coal was uneconomic and the boilers for the generating plant were run on wood.

But it was in the bigger cities that demands increased for the "isolated plant" that Edison rather ruefully realized were swamping demands for connection to any central plant. Preece, the converted British skeptic, noted the growth of individual plants in a report to the Post Office after he had toured the United States a decade later. He reported:

> In all large cities there are immense blocks of buildings fitted with lifts and used as offices. In all cases they were lighted by their own individual plants. One of the most striking features of a bird's-eye view from some lofty spot of a great city, like New York or Chicago, is the white, fleecy, wool-like jets of vapours that float gracefully away from the top of every large building. They are to be reckoned by the hundred, and every one means steam plant, lifts and electric light.

Abroad, also, it was the isolated plants that grew in favor, and Edison's incandescent lamps were soon lighting theaters in London, Berlin and Prague, breweries, paper and woolen mills in France and Germany, and a variety of factories across Europe. The House of Assembly in Melbourne and the Government buildings in Brisbane were lit by the Edison system, while the New York Company's announcement of what it or its subsidiaries were doing included the lighting of a "man-of-war" at Trieste and an "arsenal" in Spain. A new touch was added when, at the Crystal Palace Electrical Exposition of 1882, the world's first electric sign spelled out EDISON above the organ in the concert hall, a touch that was refined the following year when above the Company's pavilion at Berlin's Health Exhibition a motor-driven sign spelled out Edison's name letter by letter.

In Santiago, Chile, it was not only lighting that had been installed. Into the lighting system of the first four hotels to be equipped with electricity, a new electrical fire alarm system was built. "By this device," it was announced, "the guest is not only apprised of the danger, but is provided with sufficient light by which to escape. The cost of installation is small, and the working expenses trifling, while the arrangement is so simple that nothing can ever get out of order."

However, electric lighting and electric fire alarms were not the only benefits to be enjoyed when a house was connected to the Edison mains. Edison himself wrote of electricity in the *Boston Herald*:

> It is so easy of control, the apparatus required so inexpensive, that it can be used as a motor power for purposes innumerable. In a house it can be used to drive

miniature fans for cooling purposes, to operate a sewing machine, to pump water, to work a dumb waiter or an elevator, and for a hundred other domestic uses which now require personal labor.

The problems were slightly greater than Edison's customary ebullience suggested; but here he was thinking far ahead of the rest—toward the "all-electric house" of three decades on.

There were other ingenious ways in which he planned to use the new lamps. One involved the use of a small dynamo which could be moored in harbors and powered by wave action. The incandescent lamps, lighting up intermittently, would serve as warning lights—a nineteenth-century version of the atomic age harbor lights that need no maintenance and run almost indefinitely.

Such ideas were useful not only in themselves, but in keeping alive public interest. In this, Edison showed a flair almost as great as his genius for invention, and the records of the 1880s are spattered with ideas that would do credit to any contemporary advertising chief. There was, for instance, the Negro attendant to be found at all Edison exhibitions, his clothing wired to a lamp on the top of his helmet. His shoes had points with which he could make contact through the carpet of the exhibition stand and as he handed out leaflets his lamp miraculously lit up.

There was also the hollow square of several hundred men, each with a lamp-mounted helmet, which marched down Fifth Avenue on one occasion. The lamps were lit from a steam engine and dynamo inside the square by means of neatly hidden wires; so was the light on the tip of the baton wielded by the marshal who led the parade on a prancing horse.

Backing up these isolated publicity demonstrations there was the publication, at about ten-day intervals, of the Edison Electric Light Company's Bulletins. Initially produced for the Edison salesmen who were spreading across the United States, they gave enthusiastic accounts of new installations. In the store of F. B. Thurber & Company, wholesale grocers, it was reported, there was

a long narrow room over seventy feet in length and lighted only by windows at each end. In this room more than fifty clerks do clerical work all day. The heat from gas has proved injurious to health, and the gas light has proved injurious to eyesight. This room is now lighted by one of our isolated plants and the injurious effects of gas are entirely removed.

Not content with disparaging the characteristics of gas, the Bulletins regularly described the more lurid gas explosions, feeding their readers with gory details which by contrast underlined the safety of electricity just as the gas companies were warning against the dangers of accidental electrocution. Many unexpected virtues were discovered for electric lighting. In theaters it was claimed that acoustics were improved since "a layer of heated gases act as a screen for sound, hence the volumes of hot fumes arising from the old gas footlights obstructed and marred, to some extent, the voices of the singers." Although the "extent" was not given, it was confidently stated that in such cases electric light "benefits the ear as well as the eye." Another Bulletin announced: "The Electric Light Cures Shortsightedness," an extrapolation from a report that the heat of the gas mantles might be the cause of such trouble.

Edison did his best to keep his hands on the rapidly multiplying activities of his various companies and by the early 1880s resembled the circus juggler trying to keep a dozen balls in the air at the same time. He was actively concerned with the technical perfection of the incandescent lamp, with the mass of ancillary equipment needed for the Edison system, and in particular with improvement of the dynamo. He was naturally involved in the financial ramifications of the Edison Electric Light Company, with the companies being set up to exploit the incandescent lamp abroad, and in the complex patent situation. For the first years of the 1880s he tended to brush patent matters aside while he concentrated on other things; nevertheless, they were constantly at the back of his mind.

It was in these circumstances that Edison, the practical inventor *par excellence*, missed the significance of what was later to be recognized as one of the most important scientific discoveries of the late nineteenth century. This was the "Edison effect" as it was to be called, a phenomenon which he incorporated in a patent but which he let others exploit into the wireless and electronics industries.

In his ceaseless efforts at Menlo Park to improve the incandescent lamp still further he noticed black deposits inside a bulb with which he was experimenting, and a blue halo surrounding one of the legs of the carbon filament. With curiosity aroused, and thinking that the phenomenon might be caused by molecular bombardment, he coated the outside of a bulb with tinfoil, then connected it in series with a galvanometer and the positive terminal of the filament. A current, he found, then flowed across the gap between the hot filament and the tinfoil. Moreover if platinum foil was put inside the bulb, between the legs of the filament, then the "Edison effect" was greatly increased. He noted the fact in his diary, described it in a technical paper, and in 1883 incorporated it in a patent reporting the discovery.

Here was the small seed from which the electronics industry grew. But even the existence of electrons had not yet been discovered and it was not Edison alone who failed to see what use the "Edison effect" could be. Shortly after patenting what he saw as a voltage-regulating device, he sent specimen lamps to Preece, who read a paper on the phenomenon. But Preece neither explained it nor suggested any technical application; this was not to come until two decades later. The current, Fleming then found, consisted of the recently discovered electrons boiling off the hot filament on to the cold plate. Since the electrons were negatively charged, this effect would take place only when the plate was attached to a positive terminal of a generator. Thus when alternating current was fed into such a device only direct current would leave it.

In 1904 Fleming, working on the newly born wireless and trying to change a feeble electrical oscillation into a feeble direct current, found the first, and epoch-making use for the phenomenon noticed in the Menlo Park laboratory almost a quarter of a century earlier.

> Thinking over the subject, intensely [he wrote in his memoirs] I had in October 1904, a sudden very happy thought. I recalled to mind my experiments on the "Edison effect," and in particular my observation that the space between an incandescent carbon filament and a cold metal plate in a bulb exhausted of its air had a one-watt conductivity for electricity. Then I said to myself, "if that is the case we have here the exact implement required to rectify high-frequency oscillations." I asked my assistant, G. B. Dyke, to put up the arrangements for creating feeble high-frequency currents in a circuit and I took out of a cupboard one of my old experimental bulbs....

The experiments were a complete success. The outcome was the radio valve—known as the tube in the United States, where Lee De Forest had been doing similar work—a more efficient piece of equipment than the coherer, the magnetic detector, or the crystal detector which at the time were the only alternatives. That Edison himself failed to follow up the famous "effect" is explained by the huge amount of work involved in exploiting the incandescent lamp.

He was, moreover, by this time deeply engaged in another enterprise. It was the possibility of light, narrow-gauge electric railways, serving the seasonal wheat trade of the Great Plains for which steam would be uneconomic, that had helped to renew Edison's interest in electricity. While he had been concentrating on the search for an incandescent lamp he had continued to think of electric motive power. There had been attempts to develop this since Thomas Davenport of Brandon, Vermont, had in 1834 devised a model railway which used batteries for current. Although Davenport, like Edison, used a stationary source of power, taking the current in through one rail into a motor on the moving vehicles, and then out through the other rail, most of his successors mounted the power source on the vehicle. The difficulties which heavy batteries involved were removed with the coming of the practical dynamo and in 1879 the German firm of Siemens demonstrated at a Berlin Trade Fair a five-car train drawn by a vehicle powered with a 3-horsepower motor.

There were inevitable teething troubles with the Siemens train; not least was the lack of insulation which limited its demonstration to dry days alone. Edison was confident he could go one better. By May 1880 he had laid out from the Menlo Park laboratories a track made from street-car rails that ran for a third of a mile, first following a country road, then passing round a small hill and returning to complete a rough "U." Current was fed to the track by two dynamos and into a third one mounted on its side on a four-wheel iron truck and used as a motor. Power was then led to the driving axle of the truck by a series of friction pulleys.

"Everything," stated a report in the *Electrical World* of the train's first journey, "worked to a charm until, in starting at one end of the road, the friction gearing was brought into action too suddenly and it was wrecked. This accident demonstrated that some other method of connecting the armature with the driven axle should be arranged." The problem was solved by devising a belt drive which incorporated an idler-pulley, operated by a lever to tighten the main driving belt and send the train on its way.

Edison clearly had high hopes of his electric railway and before the end of 1880 was filing the first of numerous patents covering the system. However, it is difficult not to believe that the adventurous excitement of trying out "the train" also aroused his essentially schoolboyish enthusiasm. This is suggested by a report in the *Scientific American.*

> By invitation of Mr. Edison, representatives of this journal were present at a recent trial of this novel motor, & had the pleasure of riding, with some twelve or fourteen other passengers, at a breakneck rate up and down the grades, around sharp curves, over humps and bumps, at the rate of twenty-five to thirty miles an hour. Our experiences were sufficient to enable us to see the desirableness of a little smoother road, and to convince us that there was no lack of power in the machine.

Besides obvious enjoyment at "the humps and bumps" there was a determination to make the electric railway a practical proposition. Edison was often urged on by a genuine social motive; in this case the wish to provide not only an answer to the transport problems of the Middle West's wheat growers but also a substitute for the steam locomotive, a train which would, as the *New York Herald* (25 June) put it, be "most pleasing to the average New Yorker, whose head has ached with noise, whose eyes have been filled with dust, or whose clothes have been ruined with oil."

At first the electric locomotive was hardly taken seriously. Most railroad operators saw little reason to investigate what seemed to them little more than a money-wasting speculation. Only a handful of the general public even knew what electricity was and were hardly able to assess its potentialities. More curiously, there was ridicule from those closely linked with the development of electricity. The *Operator*, one of the leading electrical journals in the United States, had this to say of the project:

> Mr. Edison is to the front again with another idea. This time he does not offer us a new anaesthetic, a remedy for baldness, or a patent cradle rocker. He has "invented" an electric railway; but it is somewhat curious that his invention follows, with a suspicious interval, the previous announcement, in these columns and elsewhere, of a similar invention, by Dr. Werner von Siemens of Berlin.... We quote from a professedly scientific report, which adds that the great successor of Barnum realizes 70 percent of the original energy, and apparently hopes to double it—we shall not be astonished if he does....

The *Operator*, like most of the other journals which criticized Edison, was well aware that in an age of expanding technology many men were developing similar inventions and leapfrogging each other up the road to success as they translated ideas into hardware. Edison aroused suspicion because of the enthusiastic optimism which so often infected those who interviewed him. Furthermore, he tended to arouse envy; if his announcements of success were frequently premature they were usually right in the end.

With the electric railway he was not to be so lucky. He continued to improve the running of vehicles on his 1,300-yard track, he continued to patent the devices he invented for them. But the struggle to perfect a lighting system took up more and more of his time and it was only in September 1881 that he made the next step forward.

It was then that Henry Villard, President of the Northern Pacific Railroad, underwrote the building of a three-mile track at Menlo Park and agreed to build fifty miles of electric railroad once Edison had reduced operating costs to below those of steam. Before this could be achieved Villard's finances had toppled and Edison had found that the electric traction patents he had lodged were being challenged by S. D. Field, an engineer with whom it was eventually found expedient to collaborate. The result was the Electric Railway Company of America. A few months later Edison and Field installed a demonstration track, a third of a mile long, in the gallery of the Chicago Exposition. Along this track, incorporating a central rail from which the current was picked up as well as the normal outer two rails through which the current was returned, the *Judge* hauled a car-load of twenty passengers. Its work, noted the *Electrical World*, "will compare favorably with that of any of its steam rivals running over the country," and before the exposition ended, it had carried more than 26,000 passengers a total of 466 miles. "There is," the *Electrical World* reported, "solid ground for hope that in the near future we shall see it supplanting horseflesh and steam; see it drawing carriages over mountains and rivers, without clatter of hoofs or noise of snorting engines."

This was all to come, although Edison was to be only marginally involved. The Electric Railway Company failed to prosper for a variety of reasons, while Edison himself gave the simplest explanation of his failure with railroad plans to the *Electrical World*. "I could not go on with it because I had not time," he said. "I had too many other things to attend to, especially in connection with electric lighting."

When his interest revived a few years later he spoiled his case by extravagant claims. In 1891, after he had painted a glowing picture of what electric railways would do, the *Electrical Engineer* was critical:

> Mr. Edison today is unfortunately just as free and unconfined in the expression of his hopes and beliefs and opinions as when he was a humble operator; and he is just as fond of turning over in his mind all sorts of extraordinary projects as the day when he made his first invention. ... Few [reporters] are informed on electricity and mechanics, and probably still fewer of them are aware that Julian Hawthorne once said that if Mr. Edison would quit inventing and go in for fiction, he would make one of the greatest novelists this country ever saw.

It was not only Edison's claims for the electric railway that were to arouse censure. Frank Sprague, who had calculated the thickness of the Pearl Street feeders, had left Edison's employment to found his own Sprague Electric Railway & Motor Company. Edison had been free with his praise of Sprague's early efforts and of his exhibit at the Philadelphia Exposition in 1884 had declared Sprague's to be "the only true motor: the others are but dynamos turned into motors." But five years later Edison General Electric took over the Sprague Company and Sprague's name was dropped. The almost inevitable happened. "Edison was honored and decorated, a sense of his greatness was forced upon him," as Roger Burlingame has caustically but fairly commented in *Engines of Democracy*; "he accepted it with reluctance, but no man can resist such pressure. Often when he was given credit for something like electric traction which belonged, *in toto*, to others, he seems simply to have refused comment."

It was no doubt a weakness. Together with the ebullience of his own claims it throws over his achievements a darker shadow than is warranted by the facts.

Chapter Eight

Controversies, Currents, and the Phonograph

The mid-1880s saw the triumph of Edison's electric light venture, a success more important for his reputation than anything he had previously achieved. His multiplex system was a significant technological step forward but appreciated by few outside telegraphy. His phonograph, so far as it by then existed, was an ingenious novelty, its potentials so far unexploited. The carbon transmitter which had made Alexander Graham Bell's telephone a practical invention was in a different class; but the glory there had to be shared, at least equally and in a measure totally different from the division between Edison and his predecessors who had produced incandescent bulbs but failed to produce a lighting system. Between 1880 and 1890 Edison crossed that real but unidentifiable frontier which divides the famous from the celebrities.

The decade was also important in his life for other reasons. In 1887 Edison was forty, that age before which scientists and mathematicians must allegedly achieve their great work. The dictum is as facile as it is frequent, yet it is true that before the mid-1880s Edison had rarely put a foot wrong and that after, despite his successes, failure became more frequent. Perhaps, like the great men whose records justify the nothing-after-forty theory, he had burned himself out in the wild bouts of work at Newark and Menlo Park. Whatever the cause, genius began to flicker.

One reason for the subtle change can plausibly be credited to the death of his first wife, Mary Stilwell, and his remarriage some two years later to Mina Miller. Each wife was considerably younger than Edison. It was one of the few things the two had in common. Edison's first wife had, adoringly, allowed him to carry on as he wished, unkempt, unconventional, an undeniable law unto himself; the second Mrs. Edison, quite as loving, did her best to force him into the mold of the conventional great man. Hers was a deeply religious background and it is plain that from the time of his second marriage Edison had a hard task fighting for

149

his unbeliefs. He frequently won—but a good deal of his concentration, thought and energy now had to be devoted to a rearguard action against this conventionalization.

Internal family pressures were compounded by the fact that the new industry which Edison had created inevitably raised him from the ranks of the inventors into the board rooms of big business. However much he might hanker for the free-and-easy life of his younger days, he was now pulled more deeply into the wheelings and dealings of the financiers; not at the pinnacles where Morgan, Henry Villard, and the ineffable Jay Gould operated but nevertheless at a level where, for the first time, he exercised power. Not absolute power, of course; but, as Acton has implied, even a little has its effect.

The outcome of these factors—of age, of a new social environment, and of transition into a celebrity whose actions had repercussions far beyond North America—was not only apparent in diminishing signs of genius. Edison was too great a man to succumb to the blandishments of fame. Yet there now appears a subtle difference in his actions. In the earlier days the ebullient inventor tended consciously to ape his own natural characteristics, responding overexuberantly to newspaper reporters, trailing his coat with sometimes wild forecasts yet usually, in at least the stray remark, revealing that he understood the act he was playing. After the end of the 1880s, act tended to merge into actuality. Edison playing the Edison myth tended to become Edison playing himself. In the famous self-dramatized photograph showing him allegedly after five continuous days and nights at work on the phonograph, the Napoleonic stance was not entirely produced for advertisement.

From the winter of 1881 Edison had been spending more and more time in New York, less and less at Menlo Park. His wife and family were installed in a handsome New York apartment and within a few years the New Jersey house had become no more than a summer home. Here, in July 1884, Mrs. Edison fell ill with typhoid fever. At first there seemed little cause for real alarm and Edison, fighting a tough battle to gain control of the Edison Electric Light Company once more, remained in New York. At the first sign of a turn for the worse he hurried to Menlo Park and was by her side when she died on 9 August.

The effect of his wife's death on Thomas Edison was greater than might have been expected. His home and his family had seemed to occupy his attention only when by some mischance or mismanagement there was no immediate work to which he could turn his hand. But this was the way he wanted it; this was what had been necessary to keep him going. Without his wife, he was completely desolated.

The two Edison boys remained at Menlo Park. The daughter, Marion, was at boarding school in New York and for some months after the death of her mother was her father's constant companion. But Edison was now one of the most eligible widowers in the country; he was only thirty-seven and it was inevitable that his friends should compete with each other at finding him a new wife. He seems to have eyed their efforts with interest.

It was the wife of Ezra T. Gilliland, whom Edison had first met years earlier when working as a telegraph operator in Cincinnati, who finally produced the second Mrs. Edison. Gilliland was himself a minor inventor of telegraphic equipment. He had built up a successful business in Boston; he and Edison had remained in touch following their youthful friendship and in 1884 the two men were experimenting with a system of "wireless" train telegraphy founded on the principle of induction.

During the last months of 1884 and the early part of 1885 he paid many visits to the Gilliland home where Mrs. Gilliland—in Edison's opinion one of the few exceptions to his generalization that "women ... do not seem to want to get out of the beaten path"—invited a succession of young and eligible ladies. Eventually there came Mina Miller, the daughter of Lewis Miller of Akron, Ohio, who had made a fortune from farm implements and with Bishop John Vincent had founded the Chautauqua Association for adult education.

Miss Miller was just nineteen, and according to the *Akron Times* was "accomplished and serious, with a liking for charity and Sunday school work." But she had beautiful eyes and hair as well as a good figure. She had traveled in Europe and despite her youth appeared well equipped to handle the family and social affairs of a man now dining with the Pierpont Morgans of the world. As much to the point, Edison fell in love with her at first sight.

However, notwithstanding his fame and fortune the Millers did not regard him as necessarily an ideal bridegroom for their young daughter. Of course there was the age, almost the generation gap. But in addition Miss Miller already had a suitor in the orthodox son of Bishop Vincent, while Edison found it difficult to conceal the fact that if religion occupied any place in his life it filled only the smallest nook or cranny. Nevertheless, to a man capable of surviving in the jungle of Wall Street-backed invention, these were minor, and easily surmountable, impediments. With her parents' consent, Miss Miller was shortly allowed to accompany the Gillilands, Edison, and his daughter Marion, on a tour of the White Mountains of New Hampshire.

Edison had in a few earlier meetings taught Miss Miller the Morse code and now, bowling through the White Mountains, put the accomplishment to good use. He recorded in his diary:

I asked her thus in Morse code if she would marry me. The word "yes" is an easy one to send by telegraphic signals, and she sent it. If she had been obliged to speak she might have found it much harder. Nobody knew anything about our many long conversations.... If we had spoken words others would have heard them. We could use pet names without the least embarrassment, although there were three other people in the carriage.

Before the marriage the following February, Edison had bought the house that was to be his home for the rest of his life. This was Glenmont on the outskirts of West Orange, New Jersey, a town which with Orange, South Orange, and East Orange made up the four-town group of Newark suburbs. The house stood in thirteen acres of park and gardens built by a millionaire New York merchant who had failed, absconded, and left only the house to satisfy creditors. It has been described in a eulogistic life of Edison as "refreshingly independent of architectural rules" yet presenting "a wealth of fancy, which brings into view at every turn unguessed and delightful surprises. It abounds in gabled roofs, picturesque nooks and angles, carved balconies and mellow sheets of stained glass...." It was, by modern standards, excruciatingly ugly. However, with it went not only the parkland, horses, cows, and a well-filled poultry run but an acre of glass houses, and the entire contents of the house itself, furniture, library, and artistic treasures.

Shortly before his marriage Edison visited Florida with the Gillilands and, intrigued to find an area of bamboo sixty-feet high, and with the future of the incandescent lamp still in mind, took an option on a dozen acres of land, determined to build himself a winter home in the State at Fort Myers. Here he brought his new bride for a brief honeymoon before returning to Glenmont.

Menlo Park was now behind him. So was at least some of the old style of life. He still worked as single-mindedly as ever. The order of priorities was still work first, family second. But with the birth of two sons and a daughter by Mina Miller—Madeleine, Charles, and Theodore—the children began to occupy more than casual thoughts. Years later the daughter wrote:

> *My* father was the man who drew pictures for me of beautiful if slightly angular ladies, which I did my best to copy; who thought a spectacular thunderstorm or a brilliant rainbow sufficient excuse to wake us from our first sleep and bundle us out of our cribs to see it; he was the man who telephoned Mother almost every evening from the "Lab" to "send down lunch for seven—we'll be working all night"; the man who played Parchesi with us strictly according to his own rules—now *there* was an invention for you!—and who had remarkably efficient attacks of indigestion whenever there was a party. The remarkable part was that they always occurred before, and not after, the festivities.

The children grew up in a New Jersey countryside that their father was to help transform. Even before the death of his first wife, much of the laboratory's work had been moved elsewhere and the process was now completed. The laboratory where the incandescent lamp had been born fell first into disrepair and then into ruin. Meanwhile, near Glenmont, its successor was growing fast.

One day, driving out from the house, Edison had said to his new secretary, Alfred Tate, "See that valley?" Tate replied that it looked a beautiful valley. "Well," said Edison, "I'm going to make it more beautiful. I'm going to dot it with factories."

At Menlo Park he had built the prototype of the modern industrial research and development organization. He had also gone into production, of the incandescent bulb and of various less revolutionary contrivances. But, looking ahead, he wanted something more and, when his preliminary plans were reaching fulfillment, he outlined them in a letter to J. Hood Wright, a partner of J. P. Morgan, written in November 1887.

His new laboratory in the Orange Valley would soon be completed, he said. The main building would be 250 feet long and three stories high, while there would be four other 100-foot long single-story buildings. After this preamble, Edison came to the nub.

My ambition is to build up a great industrial works in the Orange Valley, starting in a small way & gradually working up: the laboratory supplying the perfected invention models, patterns, and fitting up necessary special machinery in the factory for each invention. My plan contemplates to working only that class of inventions which require but small investment and of a highly profitable nature, and also of that character that the articles are only sold to jobbers, dealers, etc. No cumbersome inventions like the Electric Light. Such a works in time would be running on thirty or forty special things of so diversified nature that the average profits would scarcely ever be varied by competition, etc.

The process was somewhat grandiloquently described by the *Scientific American* in an article which has the character of an Edison press handout.

The method of developing an invention will be as follows [it says]. Rough sketches will be submitted to model makers, who will secure from the vast supplies of material blanks for the necessary parts, or possibly completed pieces for the apparatus, and as many workmen as can be employed to advantage will be at once detailed for the work, and thus the working model will be brought out in a very short time. Any improvements necessary are then made, working drawings are prepared, the necessary patterns and castings made, and the complete, full-sized machine or apparatus is at once constructed, tested, and if it is found to fulfill the expectations of the inventor, it is removed to be duplicated elsewhere. Inventions of sufficient magnitude to warrant the venture will be launched as the bases of separate industries.

Much of this ambitious scheme was carried out. Within the next twenty years the area saw the rise of the National Phonograph Company, the Edison Business Phonograph Company, the Edison Phonograph Works, the Edison Manufacturing Company, the Edison Storage Battery Company, and many others. Edison himself, one commentator noted, "broods over the vicinity like a small electro-motor Buddha."

The move to new life at West Orange coincided with a change in Edison's attitude to those who during the previous five years had been infringing his incandescent lamp patents with impunity. Since 1880 other filament lamps had been put on the market; so had variants of the fuses, switches, and similar ancillary equipment which formed essential parts of the Edison system. Little action had been taken in the United States and Edison had come under fire from shareholders in the Light Company for not, as it was felt, protecting their interests. There were, however, two valid reasons why he had felt it best to hold his hand. In the first place all his energy was devoted to perfecting and exploiting the lighting system. Few knew better than he the huge inroads on his time that would be made once the legal machinery was put into action. When it came to protecting his interests or pushing the electric light industry onward it was for some years almost literally an either/or choice which Edison had to make.

His decision to remain outside the courts was reinforced by his skepticism about the patent law. His most critical comments were made years later, but the views on which they were based had been held since his earliest efforts to safeguard his inventions.

> I lost the German patent on the carbon telephone through the insertion of a comma which entirely changed the interpretation of the patent [he once wrote]. Another foreign patent was lost because the patent office in that country discovered that something similar had been used in Egypt 2000 B.C.—not the exact device, but something which was nearly enough like it, they claimed, to defeat my patent.

There was, he maintained, no justice to be obtained through the patent law.

> It has resolved itself into technicalities and formulas [he went on]. A case will be thrown out of one court and carried to another, it will be sent back on writs and advanced on argument, and bandied back and forth more for the exercise of legal practice than for the attainment of justice. Where an important case might be settled in a short time by the use of common sense, it is prolonged for years through the technicality of jurisprudence, the whole course of which defeats the object sought.

It was no doubt a slightly jaundiced point of view, and a rather unusual one to be held by a man who had exercised all his ingenuity to evade a law which protected, for instance, the Page and the Bell patents. It was, nevertheless, a view held strongly enough to dissuade him from going to law during the action-packed years of the early 1880s. And it was no doubt reinforced by his memory of the week in the witness box when Western Union and Atlantic & Pacific had been contesting ownership of his multiplex patents.

Nevertheless, the change came, and between 1885 and the end of the century more than 200 lawsuits were started in an effort to protect Edison's electric lighting system, slightly less than half of them involving the incandescent lamp itself and the remainder concerned with the supporting equipment. About fifty patents were involved, the most important being that covering the use of a filament in a vacuum.

On 12 May 1885 the Sawyer–Man patent of 9 January 1880 was allowed— after a five-year campaign by Edison to prevent its granting—and, as the *Electrical World* immediately foresaw "more troublous times seem to await the incandescent lamp." Eleven days later the Edison Electric Light Company began suits to enforce its patent rights against the United States Electric Lighting Company, the Consolidated Electric Light Company, and a number of others, although only those against the first two were pursued. It also announced publicly that it was prepared

> to prosecute and punish, to the full extent of the law, all makers, sellers, or users of incandescent lamps not duly authorized by it.... The public have nothing to fear from the use of the lamp as supplied by the original inventor and discoverer, whereas they render themselves liable for damages by the use of his fundamental patent for a "filament of carbon" if unauthorized by the Edison Company.

The following year the company issued a booklet warning infringers how much they might lose.

> For the purpose of enabling you to estimate your liability in the event of the Edison patents being sustained, we may state that a conservative valuation apportions a damage of $25 for each lamp in an original installation and $2.50 for each renewal lamp, independent of the fact that such decision would necessarily render your plant inoperative....

The threats of the wrath to come were not taken during the next few years as seriously as they might have been. The United States Company held numerous lighting patents, including those of Moses Farmer and Hiram Maxim, and in the complex state of the art it was always possible that they could fall back on one of them to justify their activities. As for the case against Consolidated, that fell down when the authorities in Canada voided it due to failure to comply with certain regulations; it might be a technicality but it meant the automatic withdrawal of the Edison suit in the United States.

In Britain, where there was also litigation, the judge tried to resolve the case in an unusual way. He ordered that the instructions in the Edison specification be carried out in the presence of an independent assessor, later named as Sir George Gabriel Stokes, then President of the Royal Society. Counsel, witnesses, and experts adjourned to the Edison Lamp factory at Ponders End, on the outskirts of London, and here a number of workmen were given the Edison specification, directed to make electric lamps with tarputty filaments and to connect them up to an electric current. The lamps burned, and continued to do so for about twelve hours. Judgment was given against him.

The situation, extremely depressing to the Edison interests, began to change in mid-1889. First the Canadian decision on the original Edison patent was reversed by the Canadian Minister of Agriculture, thus allowing Edison to restart the case against the United States Electric Lighting Company. Then when Consolidated, suing Edison as he had sued them, invoked the Sawyer–Man patent, that was declared invalid. In England, moreover, where Edison had taken the case to the Court of Appeal, the lower court's decision was overturned. Things were definitely going his way. The good fortune continued and before the end of 1892 the Edison interests at last appeared free to act against infringers.

Yet as the company began to obtain injunctions against others which had been infringing their patent for years, the case was unexpectedly reopened. A Henry Goebel, whose claim to have made a practical incandescent lamp before Edison had been rejected many years earlier, now renewed his attack. Although the plea was discredited, it had the effect of holding up the Edison injunctions — and at a time when the original patent had only a few years to run. In all, the cost of litigation had by 1900 reached some $2 million, a costly defense since virtually none of the money was recoverable.

The time that Edison had to spend on the multitudinous patent cases — being interviewed by lawyers, looking out old papers and reports, giving evidence in court — had to be worked into a routine quite as strenuous as that of the Menlo Park days. Seven in the morning usually saw him at the laboratory, and if work had been carried on throughout the night his first task would be to inspect what had been done. Then he would start on his own experiments and carry on until, about noon, he broke off for lunch. It was a working lunch since he would receive visitors or confer with any of his staff who had particular problems to discuss with him.

At one o'clock he was back at his bench again, checking up on what his assistants had done and advising them as to procedure [according to one of his old friends,

John F. O'Hagan]. There were no independent experiments in the Edison laboratory. "The Old Man" was the director-general, and his ideas prevailed, as many
a new assistant soon learned after he attempted to digress from the daily written
instructions which were given him.

To visitors, and there were many of them, Edison had by now become even
more of a father figure, scolding his men with a bluff heartiness they would accept,
turning to a bench to demonstrate how a job should really be done and asking,
in laboratory or workshop, the probing questions which revealed him as a
master of whatever craft they were practicing. Over his black suit he would usually be wearing a long smock of check gingham, buttoning close up to the chin,
reaching to his heels, and protecting his clothes from acids, oil, and dust. "Edison
in his working gown," observed one visitor, "is a sight to laugh at: until you
remember that it is Edison."

He was still a master of publicity. Almost anyone would be shown round
the West Orange works with justified pride. Then Edison would sit them down,
shoot a long string of questions, cock his head at an angle and cup his hand
over his bad ear as he waited for the answers. "There is something quizzical
in his face while he is waiting to learn your business," said one visitor. "If it
happens to be newspaper business, there is not a man living who will make
you more at home in two minutes than will he."

Edison's ability to make maximum use of the Press has sometimes been held
against him, although the difference between his attempts and those of most
competitors was one of success rather than of effort. In the early days it stemmed
largely from his own bubbling enthusiasm. Then, as the benefits of publicity
became obvious, the attempts became more studied. The natural enthusiasm
is replaced by the formalized insincerities of the publicity handout and by the
end of the 1880s there is the famous posed portrait of the Napoleonic Edison
after his renewed work on the phonograph.

The most notorious campaign in which he used the Press to propagate his
own ideas was that against alternating current. It was a campaign waged by
both sides with few holds barred, but it was one in which Edison had, for a
complex mixture of reasons, backed the wrong horse.

Until the later 1880s, his predominance as the man who could turn electrical
ideas into practical ironmongery remained virtually unchallenged. Then, in
1887, there appeared on the horizon the possibility of using alternating rather
than direct current, a subject on which he was to remain as stubbornly conserva-
tive as those who had ridiculed the idea that electricity could displace gas.

Edison's lighting system, like those of his rivals, had been built up to exploit
current which flows in one way, from the generator to the user and then back
to the generator. Electricity that can be distributed in this way is for practical

purposes limited to about 250 volts, the highest that will not burn out incandescent filaments and the highest considered safe for domestic supply. This meant that the direct current system had one built-in disadvantage, since such voltages could be transmitted economically for only limited distances. In the early days of the electric light this was not specially important. Edison's Pearl Street station had been strategically situated in the center of a densely populated area, and so had the stations which followed it. The situation began to change as the demand for electricity in less populated areas started to increase. Was it not possible, it was asked, to satisfy the demand by raising the voltage, transmitting it comparatively cheaply at the high voltage, and then lowering the voltage before the current was taken into houses or factories?

This was impracticable with direct current. But what about alternating current, generated in such a way that it flows in one direction, builds up to a peak, then changes direction, builds up to a peak and then changes back, the change of flow taking place many times a second? There had been many attempts to introduce alternating current systems in the early 1880s but all had met with difficulties. But this had been true of the incandescent lamp and it was not fear of development problems that produced Edison's conservatively stubborn reaction. He had built up a system, did not want to change it and was apparently slow to grasp the economies which an alternating current system would make possible. But there was also the human factor. It seems likely that Edison might have held out less dogmatically if the system had not been developed by his American rival, George Westinghouse, on technical foundations laid by Nikola Tesla, born in Croatia, then part of Austria-Hungary.

Tesla had been "discovered" in France by Charles Batchelor and had been persuaded to visit Edison in the United States. He arrived bearing a letter from Batchelor to Edison which said: "I know two great men and you are one of them; the other is this young man." That in itself may not have endeared Tesla to Edison, quite apart from the fundamental differences between them.

When they met Edison was world famous, wealthy, and already a symbol of success. Tesla, only a few years younger, had so far accomplished little. Edison was above all a trial-and-error man, always anxious to test for himself whether an idea would work. Tesla typified the man of theory. There was also the question of alternating current which Tesla already believed would one day allow wider and more economical distribution of electricity. Edison brushed aside such ideas and refused to take them seriously. Nevertheless, he hired Tesla.

The break came after less than a year. Tesla had one day been discussing a number of potential improvements in the dynamo and Edison had casually remarked, "There's fifty thousand dollars in it for you if you can do it." Within a few months Tesla had worked out the improvements, incorporated them in a dynamo and had it successfully tested. He then asked Edison for the fifty thousand. The answer was "Tesla, you don't understand our American humor." Tesla resigned at once. The Austrian was a touchy and somewhat eccentric character, Edison was not mean, whatever else can be argued against him, and it seems likely that the unfortunate incident, with the repercussions which it was to have on Edison's future, was the simple result of poor communication. Nevertheless, Tesla continued to feel strongly. In 1912, more than two decades later, he was offered the Nobel Prize for Physics. It would have brought him both acclaim and badly needed cash. But it was to be a joint award with Thomas Alva Edison. Tesla refused, and the Prize went to the Swedish physicist, Nils Gustaf Dalén.

After leaving Edison, Tesla gained the support of George Westinghouse and eventually succeeded in making the use of alternating current a practical proposition. The motors which he developed were simpler and more flexible than the direct current machines, while his transformers solved the problems inherent in transmitting electricity over long distances, stepping up the generated current before it was passed on to the transmission lines and then decreasing it before it was led into homes or factories.

Edison was slow to take the new threat seriously. When he did, "the battle of the currents" was fought with the gloves off. There could be little doubt about the economic advantages of alternating current and Edison eventually realized that a battle fought on economic grounds would be lost. There was, however, another possibility. In the last decades of the nineteenth century electricity was looked upon by the general public much as nuclear power is considered today: while it could be of immense benefit to mankind, it was also a killer and one about which laymen still knew very little. It was thus easy to fuddle the public mind and to claim that the dangers of high voltages—which though real could be guarded against—were those of alternating current itself. In the words of Harold Passer's study of *The Electrical Manufacturers*, the Edison company now "decided to compete outside the area traditionally identified with commercial rivalry.... As the foundation stone of his extra-market competition with alternating current the Edison company chose public safety."

Every accident that could rightly or wrongly be attributed to alternating current was publicized by the direct current party. Claims and counterclaims which owed little to science and much to dramatic copywriting were published almost daily. Among the highlights of the Edison campaign was "A Warning," bound in red and recounting in great detail the alleged dangers of alternating current. Almost incredibly, its use was described as an *ignis fatuus*, the very phrase which had been employed by Preece to ridicule Edison's idea of subdividing the electric light. But Freud was only then setting up his practice in Vienna

159

and no one remarked on the fact that Edison had abandoned his position in the vanguard for his old opponent's position as a tough rearguard fighter.

In "Dangers of Electrical Lighting," an article in the *North American Review*, Edison lambasted the idea of using alternating current for any purpose:

> The electric lighting company with which I am connected purchased some time ago the patents for a complete alternating system and my protest against this action can be found upon its minute book. Up to the present I have succeeded in inducing them not to offer this system to the public, nor will they do so with my consent.

Westinghouse replied in support of alternating current.

However, the battle was also being waged outside the magazines. There was a good deal of political lobbying, the Edison interests supporting a proposed law to limit electric circuits to 800 volts, thereby preventing the effective use of alternating current but leaving their own circuits untouched. In the summer of 1888 Westinghouse thought of legal action. On 11 July he wrote:

> It is a matter worthy of very serious consideration as to whether or not we could not proceed against the directors of the Edison Company, Johnson and others, for conspiracy under the laws of New York; for their recklessness, and you might say, criminal course should in some way be brought to an end.

Strong emotions were aroused and Edison, a naturally affable man, asked to visit Westinghouse by Villard, a mutual friend, replied uncharacteristically:

> I'm very well aware of his resources and plant, and his methods of doing business lately are such that the man has gone crazy over sudden accession of wealth or something unknown to me, and is flying a kite that will land him in the mud sooner or later.

The alternating current interests eventually triumphed, but only after a dramatic coup by their opponents. The direct current party first carried out a gruesome promotional campaign, conceived by Insull, Johnson, and Edison and carried out by Harold P. Brown, a former Edison laboratory assistant, for the use of the electric chair as a method of executing criminals. As part of a complex plot, Brown had in May 1889 bought three of Westinghouse's alternating current motors without giving Westinghouse any idea that they were to be resold to the prison authorities. A year later it was announced that future executions in Auburn State prison, Sing Sing, and Clinton would be carried out by electrocution and on 6 August 1890 William Kemmler was electrocuted for murder in Auburn. He died by alternating current and in the minds of large numbers this became synonymous with death.

The setback was only temporary and during the next few years alternating current slowly began to take over. As late as 1903, Edison himself was still protesting that he "could not understand why everyone was going mad on alternating currents" but long before this the Edison Companies had adopted it for central station generation and transmission, while retaining direct current for local distribution networks.

The change had come during the closing stages of a corporate transformation which had begun in the early 1880s and was complete before the end of the century. The setting up in 1881 and 1882 of separate companies to supply equipment for Pearl Street and its successors had been followed by a campaign which eventually gave Edison control once more of the Edison Electric Light Company, thus enabling him, as he wrote in the *Electrical World*, of 4 August 1883, to "take hold and push the system better than anyone else." Success had barely been achieved, however, when "the system" expanded into an industry so big that its financial and administrative problems were—like those of the telegraph industry when he had entered it—best solved by lawyers, financiers, and promoters rather than by inventors. He was canny enough to sense the change of emphasis and encouraged a series of amalgamations which resulted in the formation of the Edison General Electric Company in 1888. "I have been under a desperate strain for money for twenty-two years," he wrote the following year to Henry Villard, president of the new company, "and when I sold out, one of the greatest inducements was the sum of cash received, which I thought I could always have on hand, so as to free my mind from financial stress, and thus enable me to go ahead in the technical field."

With the creation of the Edison General Electric Company Edison surrendered his firm control of a large part of the new industry he had helped to found. What he received in return was money and time to expend on the business of inventing in general and, in particular, on development of the phonograph, search for "the talkies" on which he had already speculated, and perfection of the ore-crushing mills on which he was to lose a fortune.

In all these enterprises he was aided by able and imaginative assistants and profited from the experiences of other men working in the same field. It is easy, therefore, to minimize the part played by Edison himself and to judge him as the coordinator of a team rather than as an individual inventor. To some extent this is justified; the earlier role of the inventor was already being outdated by the quick trot of technology. But if success demanded the coordination of many crafts and the exploitation of different and unlinked researches, this was the very task for which Edison had built up his "inventions factory." Thus his role was now different from that of his early telegraph days: that of conductor commanding an orchestra rather than that of star soloist.

161

The spur which drove Edison back to the phonograph was the challenge raised when Graham Bell prepared to put on the market an improved machine beside which Edison's original instrument would be shown up as nothing more than a novelty.

Bell's father-in-law had been a shareholder in the Edison Phonograph Company and Bell himself showed a keen interest in the early machines, directly concerned as they were with his studies of speech reproduction. In 1880 he won the Volta Prize of $10,000, spent most of it in creating his own sound laboratory in Washington, and by the following year had developed his own phonograph. He acknowledged his predecessor and a recording which he lodged with the authorities to substantiate any future claims ended with the words: "I am a graphophone and my mother was a phonograph."

Bell's cousin, Chichester Bell, and the instrument maker Charles Sumner Tainter, were soon co-opted and by 1887 they had developed Bell's prototype into an electric-powered machine which used waxed cardboard instead of Edison's tinfoil, and a loosely mounted stylus in place of his more rigid recording device. They then approached Edison and suggested that the new machine be marketed under their joint names.

According to his secretary, Alfred Tate, Edison had by this time almost forgotten his potentially important invention of a decade earlier.

> But the visit of those gentlemen from Washington revivified his memory and shocked him into action. He refused to entertain their proposals. He would not deal with them on any terms. The idea of incorporating under his name anything that did not represent the product of his own labors impressed him as a fraudulent device designed only to achieve wealth, and he wanted no wealth that he himself did not create.

To Tate he complained, revealing a touch of paranoia: "Those fellows are a bunch of pirates. They are trying to steal my invention. I'm going ahead now to improve the phonograph and I'll show them that they can't get away with it."

There were formidable problems ahead and Roland Gelatt, the historian of the gramophone, has described one appalling failure which happened while Edison was still trying to raise interest in his machine. Bankers from the investment house of J. & W. Seligman arrived at the Orange works for a demonstration and Edison began by dictating a short message.

> He then lowered the reproducing stylus into place and prepared to let the phonograph sell itself to his assembled guests. But instead of parroting the words he had just spoken, the phonograph emitted nothing more than an ugly hiss. Was it showing its contempt for the leaders of finance? Edison made some small

adjustments, inserted a fresh cylinder, and dictated another letter—with the same humiliating result. After some further abortive tries, the Seligman entourage took their leave, promising to return when Edison had the instrument in working order. The defect was quickly repaired, but the Seligman people never paid a second visit.

It was Edison himself who discovered that one of his assistants, trying to be helpful, had fitted a new playing stylus on the machine. Unfortunately it was of the wrong kind.

Throughout the spring of 1888 he continued to cope with the machine's teething troubles and in the summer he publicly announced his new and improved phonograph. Perfecting it had been a long hard job; it was pardonable that he should have himself photographed, allegedly at the end of five days and nights of work, in a portrait that was an advertisement manager's delight.

His latest brainchild employed the floating stylus of the graphophone, was powered by an electric motor, and used wax as the recording medium. In one way, however, it took a leaf out of Bell's book: the Edison cylinders—or phonograms as they were called—consisted not of wax-covered cardboard but of solid wax. This meant that when one recording had become too worn it could be shaved off and another recording made on the new surface.

The new machine gave far better reproduction than its predecessor of a decade before and in England Colonel Gouraud, having brought a sample from West Orange to "Little Menlo," his home in a London suburb, was quick to exploit its possibilities. He soon received from Edison not a letter but a phonogram, an event which he successfully recorded in a letter to *The Times*.

> At five minutes past two o'clock precisely, I and my family were enjoying the at once unprecedented and astounding experience of listening to Mr. Edison's familiar and unmistakable tones here in England—more than 3,000 miles from where he had spoken, and exactly ten days after, the voice, meanwhile, having voyaged across the Atlantic Ocean.

From now onward he ensured that *The Times* was informed almost monthly of the modifications and improvements made to the apparatus—better reproduction of the more awkward tones, development of a "speaking trumpet" which greatly increased the range, a treadle motor which could be used if the electric motor broke down, and even a "water motor."

He also encouraged persons of note to attend demonstrations in the hope that without further prompting they would spread the fame of Edison throughout London. One was the Duke of Cambridge, that extraordinary relic from the Crimean War, his head still as empty of science as when he had been born. Shown the phonograph at a London dinner, the Duke was persuaded to say a few words into it, doubted what the result would be, and on hearing his own voice collapsed into a chair with the cry: "The Devil's in it."

Cardinal Manning was induced to receive a young lady with Edison's phonograph. He told her after a demonstration: "Ah, young lady, if you had lived a century ago, you would have been burned as a witch." Then he told her for posterity: "I trust no word of mine, written or spoken, will do harm when I am dead."

Augustus Henry Lane Fox Pitt-Rivers, General, anthropologist, and Britain's first inspector of ancient monuments, quickly saw a specialized use for the phonograph. With it, he said, "characteristic specimens of even the most barbarous music may be collected, and the travelers provided with one of them may be able to trace the gradual variation and changes of style that are found in any given region."

And in 1890 the *Scientific American*'s "startling possiblility of the voices of the dead being re-heard through this device" was turned from forecast to fact. In April 1889 Gouraud had recorded the voice of Robert Browning and the following year, after the poet's death, a small number of his friends were invited to the Colonel's home. Here a precious wax cylinder was taken from the cotton where it had been kept and put on the machine. As the cylinder began to rotate Gouraud was heard telling Edison that Browining's voice would follow his own. Then the poet asked: "Ready?"

Gouraud had apparently nodded and Browning began to recite his own poem, "How they brought the Good News from Ghent to Aix."

> I sprang to the stirrup, and Joris, and he;
> I galloped, Dirck galloped, we galloped all three.

Then he faltered. "I forget it," he said. He was prompted, continued for a few more lines, then said: "I am exceedingly sorry that I can't remember my own verses: but one thing that I shall remember all my life is the astonishing sensation produced by your wonderful invention."

In the United States the same kind of inspired publicity helped to push up sales. Eleanora Duse, possibly the most famous actress of her day, as Bernhardt had been of hers, was pleased to speak into the magic machine. So was Henry Stanley the explorer. Still the essential reporter at heart, he asked Edison: "If it were possible for you to hear the voice of any man whose name is known in the history of the world, whose voice would you prefer to hear?" Edison, answering "Napoleon's," was met with Stanley's: "No, no, I should like to hear the voice of our Saviour." Taken aback Edison replied—or is alleged by Alfred Tate to have replied—"Oh, well, you know—I like a hustler."

Publicity on both sides of the Atlantic sustained interest in the phonograph for the more serious, nonentertainment purposes that had always been at the back of Edison's mind in developing it. In the United States the business applications went ahead. From New York it was reported that "heads of firms, and confidential clerks now talk their letters to the phonograph, which redictates them to the typewriter." And a phonograph reporting company offered to record law trials, conventions, and meetings more cheaply and with greater accuracy than the conventional shorthand reporter.

However, it was soon evident that the main use of the new machine would be to record music. Large profits were in prospect, the Bell–Tainter machine was being efficiently marketed by the American Graphophone Company, and it was clear that Edison had a fight on his hands. The purely commercial opposition was compounded by the fact that the Graphophone Company appeared eager to sue Edison for patent infringement since his machine, like theirs, not only engraved a wax record but used the floating stylus. Edison, however, was unable to take quick effective counteraction despite his patent of 1878 since this described a method of "embossing or indenting" while the graphophone recorded by a basically different method: it engraved. All seemed set for a long and expensive court case. It was averted, almost at the last minute of the last hour, by the intervention of Jesse H. Lippincott, a Pittsburg businessman. Lippincott first took over the rights of the Graphophone Company; then he bought Edison's patent rights in the phonograph, although leaving Edison free to manufacture the machine.

Notwithstanding this amiable solution, argument about the relative importance of Edison, and of Bell and Tainter, in the birth of the gramophone has continued, Edison's tinfoil machine being dismissed as not serious. Here Gelatt has the last, and best, word.

> The partisans of Edison [he sagely sums up] might remember that any patent invites improvements and modifications, and that if the patentee does not make them others undoubtedly will. And the partisans of Bell and Tainter might bear in mind that the graphophone was originally introduced as a refinement of the tinfoil phonograph, with full credit going to Edison for the basic conception.

Edison had claimed a decade earlier that the phonograph could be used for a large variety of tasks and he now set about exploiting the potentialities of his vastly improved machine. One was the talking doll, which had been forecast in his long article in 1878 on the future of the phonograph: "A doll which may speak, sing, cry or laugh, may be safely promised our children for the Christmas holidays ensuing," he had then written. "Every species of animal or mechanical toy—such as locomotives, etc.—may be supplied with their natural and characteristic sounds." Even the massive expansion of a decade later did not include roaring toy lions or hooting toy trains. Yet half of the 500 workers at the Orange factory were soon at work equipping dolls imported from Europe with "the necessary vocal apparatus."

The tin body contained a miniature phonographic cylinder which could be rotated by a spring mechanism wound up with a key. When this was done and the mechanism operated by turning a small crank, a reproducing stylus moved over the cylinder and the recorded words issued through the perforated tin front of the toy. The Orange factory could equip about 500 dolls a day and buyers had a choice of the words, mainly nursery rhymes, recorded by girl workers. "A large number of these girls are continually doing the work," the *Scientific American* recorded. "Each one has a stall to herself, and the jangle produced by a number of girls simultaneously repeating 'Mary had a little lamb,' 'Jack and Jill,' 'Little Bo-Peep,' and other interesting stories is beyond description."

Not all the dolls were restricted to nursery rhymes. At Christmas, 1889, Edison presented one to the young Archduchess Elizabeth, daughter of the late Crown Prince of Austria. It recited, among other things, a poem written for the Emperor by a relative of the little girl.

Although Edison lost no time in exploiting the toy market, the new machine already reproduced the human voice with an accuracy far greater than that of his earlier phonograph. It was thus lifted to a new level of importance so that not only Mr. Gladstone, but even Queen Victoria herself graciously recorded a message on the phonograms which Edison sent to Colonel Gouraud in London. Henry Irving, James Knowlton, editor of the *Nineteenth Century*, Sir John Fowler the builder of the Forth Bridge, and the Earl of Aberdeen, were others who recorded messages of goodwill. All this was fine publicity for a display of the machines at the Crystal Palace where, before a program of music, listeners heard "The Phonograph's Salutation," a poem from an American minister which ended:

> Hail! English shores, and homes, and marts of peace,
> New trophies, Gouraud, yet are to be won.
> May "sweetness, light," and brotherhood increase:
> I am the latest born of Edison.

Two years afterward a phonograph was demonstrated at the Court of the German Emperor in Berlin, and Bismarck proposed that concealed machines should be used to record the discussions at diplomatic conferences. This would, he reasoned, "be a dangerous thing for diplomats, and also a good thing, as they would be forced to tell the truth." He was also, reported *The Times*, "much struck with the uses to which the phonograph might be turned to convey instructions to subordinates without writing and to learn requests and reports of officials and other persons without seeing them."

In the United States more imaginative schemes were apparently discussed. Edison had, it was reported,

> entered into a contract to furnish a monster phonograph to be placed in the interior of Bartholdi's colossal statue of Liberty, now being erected in New York harbor, the sounds from which will be conveyed from one end of the city to the other. It will also be available for communication with foreign and other vessels in the harbor and vicinity. At all the lighthouses and signal stations on the American coast Mr. Edison is commissioned to construct *Vocalized Foghorns*, to warn merchantmen, coasters, and other vessels from the many dangerous shoals, and other obstructions found on the great highway of waters—and which may be distinctly heard above the roar of winds and waves for from four to six miles from the station.

However, it was in the reproduction of music that the new phonograph showed its superiority over the machine of the 1870s. An early demonstration was given during the Handel Festival at the Crystal Palace in 1888. The report from W. K. L. Dickson, an Edison employee, is hardly unbiased but does suggest the impression made.

> A gigantic horn, placed in the press gallery of the Crystal Palace concert room, gathered up the majestic harmonies of the composer, in the several vocal and instrumental settings [he wrote]. Four thousand voices, a thunderous organ and a mammoth orchestra combined in the exposition of Handel's "Israel in Egypt," and this titanic volume of sound, with its finer contrasts of light and shade, was reproduced by the phonograph in a manner little short of the miraculous.

It was on music that the Edison company concentrated after the Lippincott take-over and the settlement of the dispute with the Bell–Tainter group, and for many years the Orange factory was hard pressed to keep up with orders. Edison was always trying to improve the quality of the recordings, and no detail escaped him. An example is found in an account by a young chemist, M. A. Rosanoff, who was handed a chunk of the wax from which the "master" cylinders were being made.

> When an extra-loud sound occurs in a song—you know, when an Eyetalian has suddenly fallen in love or somep'n—the recorder needle gives a jump, and then a tiny bit of the wax is chipped out ... [Rosanoff was told by Edison]. This wax was worked out for me by a fellow named Aylsworth, who used to be my chemist here, and it's a pretty good wax. But it's got to be softened a bit to be *real* good, and I'm sure, with all your college training—in Paris and everywhere else—you can do it in no time.

Rosanoff eventually did do it, but only after eighteen months' hard work.

The first serious opposition to Edison's cylinders came from flat records which had become practicable after Emile Berliner had developed a system in which the needle of the machine moved from side to side rather than up and down—the "hill and dale" system used by both Edison and Bell. The first Berliner disks ran for only one or one and a half minutes but by the early 1900s this had been increased to four minutes: this was double the playing time of Edison's phonograms, and flat records began to make inroads on the cylinder market.

As one countermove, Edison opened a New York studio and began to record members of the Metropolitan Opera Company. The decision was a major innovation since the music on which the Edison company had so far concentrated reflected Edison's own opinions and preferences. These were revealed when he met Sousa. "The public is very primitive in its tastes," Edison noted. "My object is to reach the greater number with the most wholesome kind of appeal." As to what wholesomeness was, he had definite views.

> A few people like the most advanced music, very very few [he confided]. The Debussy fanatic thinks that because he likes Debussy, there must, of course, be thousands and thousands who do. He would be amazed if he knew on what a little musical island he is standing. You could hardly see it on the great musical map of the world. All the world wants music; but it does not want Debussy; nor does it want complicated operatic arias.

And it was possibly of Debussy that Edison was speaking when he told Sousa:

> I used to reverse some tunes we had upon the records and the results were surprising. We played them backwards and some of the reversed tunes were far more interesting and charming than the originals.

None of Edison's efforts to break into the middle and highbrow phonograph market was particularly successful—partly for technical reasons but also, very probably, because his heart was not in it—and in 1908 he tried another tack in an effort to keep business moving. This was election year, and the Republican and Democratic candidates, William Howard Taft and William Jennings Bryan, were both induced to make recordings thus enabling Edison's publicity department to announce: "No matter how the November elections may result, we shall have records by the new President. This makes history. It indicates progress."

Another, and equally enterprising, touch was the sending of a phonograph and a number of blank cylinders to Tolstoi. Edison had originally thought of the phonograph as an office aid, and now heard that Tolstoi was answering about twenty-five letters daily. What more useful than a phonograph to help him? The Russian writer did, in fact, use the machine for a short while, and some of his replies have survived. But he found its operation too complicated and abandoned it after a few months.

Just as Edison had been reluctant to give up direct for alternating current, so was he reluctant to abandon the cylinder for the flat record. Thus in 1909, when he introduced a phonogram which had 200 lines to the inch instead of 100, and therefore ran for four minutes instead of two, it was still a cylinder. Quality was excellent, but not excellent enough to convince the public that they should buy cylinders instead of flat disks. Not until 1913 did Edison finally capitulate. The Edison disk phonograph then introduced rapidly began to win back the market and its flat records were being turned out by the ten thousand when in 1914 the European war suddenly cut off supplies of essential raw materials and pushed Edison into one of his most successful improvisations.

In spite of the fortune which recorded music brought him he never lost sight of one phonograph application which had been high on his list of priorities in 1877. He believed that it could be a revolutionary piece of office equipment, although if the typist missed a few words from a recorded phonogram there was trouble. In practice, she had to stop typing, move the phonograph needle back a short distance, then wait until the words were heard a second time. The addition of a lever to move the needle was an improvement but she still had to turn from her machine. The delay was finally eliminated by a device which Edison called the transophone, operated by an additional key at the end of the typewriter keyboard. If the typist missed a few words all she had to do was press the transophone key to operate a circuit which lifted the needle from the phonogram, automatically set it back a distance that could be chosen to cover a few words or many, and then allowed it to repeat the message from that point.

There was also the telescribe, a means of coupling up phonograph with telephone in such a way that a conversation could be automatically recorded. It is perhaps significant that one of the uses of the device was described by the *Scientific American* as "the transmission of late advertising copy by telephone where the exact words must be secured beyond question."

Chapter Nine

In Search of
the Talkies

Throughout the long struggle to retain a hold in the phonograph market Edison was involved in at least one major enterprise and usually more than one. The first of these was the development of the movie camera, an operation whose history has been subject to more revision over the years than has any other Edison venture.

"In the year 1887," wrote W. K. L. Dickson when describing the early history, Mr. Edison found himself in possession of one of those breathing spells which relieve the tension of inventive thought." The breathing spell, according to Dickson, quite fortuitously provided the opportunity for Edison to start work on what was to become the world's first movie camera. Since Dickson was largely responsible for development of the Orange laboratories' first moving film, and since Edison himself approved the statement, this story of how the picture industry came into existence for long remained unchallenged. Yet there is no doubt about Dickson's amazing flights of fancy and as Burlingame has underlined, "to weed out the facts from Dickson's accounts of what he thought were about to become facts is a task that has greatly complicated research into the history of the cinema." Recent investigation, notably by Gordon Hendricks of the University of California, and by Kevin MacDonnell, the biographer of Eadweard Muybridge—"the man who invented the moving picture"—tells a different and undoubtedly more accurate story.

As with other Edison inventions, it is an unfair oversimplification to imply that in developing moving films he deliberately pirated the work of other men. He did, however, often fail to admit that he was but one of a host of workers moving forward on a broad front and that cross-fertilization of ideas was inevitable. And he did take it for granted that he should be given credit for the results of men working under him as a team, a custom quite acceptable in science, where contributions by team members are usual, but more debatable in the technological development of a basic idea. These factors, which tend to confuse the unraveling of who did what, were also compounded by Edison's at times rather casual story telling, and his habit of adding contradictory details to successive listeners. It seems doubtful if he worried much about such things. The past was water under the bridge, and the essential thing was to get on with more inventions. Historians and the despised academics could worry about the facts.

In the case of the movie camera, he was later to say "the idea occurred to me that it was possible to devise an instrument which should do for the eye what the phonograph does for the ear, and that by a combination of the two, all motion and sound could be recorded and reproduced simultaneously."

By the later 1880s this was becoming a subject for discussion among scientific men, but Edison had in fact thought about it more than a decade earlier. William Bishop's account of "A Night with Edison," published in November 1878, reveals that even at that early date Edison was considering "the combination proposed by Dr. Phipson of the phonograph and kinctoscope [sic], by which a phonographic image is to move and seem to talk."

Nothing is known of Edison's Dr. Phipson, but earlier in 1878 the Englishman Wordsworth Donisthorpe had proposed in *Nature* that the phonograph could be combined with a strip of photographs showing people in motion, taken in succession and lit by electric sparks.

> I think it will be admitted [he went on] that by this means a drama acted by daylight or magnesium light may be recorded and reenacted on the screen or sheet of a magic lantern, and with the assistance of the phonograph the dialogue may be repeated in the very voices of the actors.

Nothing appears to have come of the proposal although a crude method of simulating movement where none existed had been known since a Belgian had invented the "phenakistoscope" in 1832, a device followed in the 1840s by the "zöetrope." Both made use of the fact that when an image is impressed upon the retina of the human eye, that image remains after the object that caused it has been removed—for a tenth to a seventh of a second according to the individual. The devices that made use of the phenomenon, developed before photography was born or while it was still in its infancy, employed a series of drawings which showed the successive stages of movement of, for instance, a galloping horse or a human dancer. The drawings were attached to the sides of a cylinder and viewed through a vertical slit as the cylinder was rotated. The viewer saw a drawing whose image persisted for a fraction of a second until the next drawing came into his line of vision. This image was only slightly different, as were those which followed, and the illusion of movement was complete.

The birth of photography might be thought to have altered all this. In fact, the technical limitations imposed first by the wet-plate process and then by the dry-plate, which despite its fewer disadvantages also demanded glass negatives, were still formidable in the 1880s.

However, as early as 1880 there had been one famous attempt to produce "moving pictures." It had been made by the brilliant if eccentric Englishman, Eadweard Muybridge, on behalf of the California senator Leland Stanford who had maintained, in the face of accepted beliefs, that a trotting horse left the ground entirely at regular stages of its movement. Muybridge set up a succession of cameras alongside a track and arranged that they should be operated by strings which were broken as the horse passed in front of each camera. The prints made from the negative were astounding since, as one viewer noted,

> the most careless observer of these figures will not fail to notice that the conventional figure of a trotting horse in motion does not appear in any of them, nor anything like it. Before these pictures were taken no artist would have dared to portray a horse as a horse really is when in motion, even if it had been possible for the unaided eye to detect his real attitude.

Muybridge photographed other animals and birds in the same way, made prints from the negatives, mounted them on an adapted form of zöetrope, and then projected them on a screen by means of what he called a "zoopraxiscope."

"Nothing was wanting but the clatter of hoofs upon the turf, and an occasional breath of steam from the nostrils, to make the spectator believe that he had before him genuine flesh-and-blood steeds," the *Scientific American* reported. However, whatever their scientific value, the moving pictures had damning disadvantages when considered as entertainment. To take photographs of a trotting horse for a complete minute would require, so fast did the images have to succeed each other, no less than 720 cameras. Furthermore, the animal was in each case in the center of the photograph, so that while its legs were in constant movement it was apparently making no progress, even though the background scenery seemed to be moving rapidly.

By 1888 Muybridge was giving demonstrations of his zoopraxiscope throughout the United States and on 25 February he came to West Orange. It is not certain whether Edison attended the demonstration, although it is very likely. What is certain is that two days later Muybridge visited Edison's laboratory and, in the words of his unique book *Animals in Motion*, "consulted with Mr. Thomas A. Edison as to the practicability of using [the zoopraxiscope] in association with the phonograph."

Edison, in a statement made in the *New York World*, agreed. According to Muybridge's biographer, "Edison was enthusiastic and offered to record the voices of people such as Edwin Booth and Lilian Russell, while Muybridge made moving pictures of their gestures and expressions." Eventually the idea was dropped, apparently because Muybridge felt that the phonograph lacked the volume necessary for use with a large audience.

However, the idea of moving pictures—with or without sound—remained in Edison's mind and throughout 1888 a long series of experiments was carried out at West Orange under the supervision of William Kennedy-Laurie Dickson. Allegedly a descendant of Annie Laurie of the song, Dickson had emigrated to America specifically to join Edison's staff and for a while had been superintendent of the testing and experimental department of the Goerck Street works where the Jumbo dynamos had initially been built.

The first attempts to improve on Muybridge involved making a series of tiny negatives, sometimes only a sixteenth of an inch square, on a sensitized cylinder that was moved between each exposure. The result was a spiral of small pictures. Viewed under a microscope as the cylinder was revolved they gave the same illusion of movement as did the zöetrope, although in this case the figures had been photographed rather than drawn. Some of the first film performances were as primitive as the equipment which recorded them. One of the early "stars" was John Ott, the young assistant Edison had employed since the Newark days. "I had a white cloth wound around me," Ott once said in describing an early appearance, "and then a little belt to tie it in around the waist so as not to make it too baggy—look like a balloon—and then tied around the head; and then I made a monkey of myself."

In the fall of 1888 Edison filed a caveat in the U.S. Patent Office covering his device, describing it as an instrument "which does for the eye what the phonograph does for the ear," and christening the instrument the "kinetoscope." One problem had quickly become apparent. The various wet-plate photographic processes then available were not sensitive enough to give an unblurred record of men or animals in motion. Yet the dry-plate method used by Muybridge, while being fast enough for the purpose, had a compensating disadvantage: when the minute negatives were increased in size by viewing, the silver particles held in suspension in the emulsion produced the unacceptable, almost stippled effect which photographers call "graininess."

While the West Orange laboratory was trying to cope with this and allied problems, others were also at work on what was to become "the movies." Early in 1889 Friese-Greene in Britain used a ribbon of sensitized material introduced in the United States by George Eastman a few years earlier. Pictures could be taken on this, thus abolishing the need for glass plates which had previously been necessary. He also lodged in June of the same year a provisional application for a patent covering a camera to use the film, although its speed of planned operation was so slow that many experts cannot see it as the genuine ancestor of the movie camera. In the United States, George Eastman was about to market his film for use in still cameras and, illustrative of the way in which research was moving forward at the same time in different places, Edison had on 30 May 1889, a few days before Friese-Greene's patent application, written to Eastman for a supply of his new film. "That's it—we've got it," Edison said when it arrived. "Now work like Hell."

Simultaneously with his application to the British Patent Office, Friese-Greene wrote to Edison. He had owned an Edison phonograph for more than a year and had photographed a man singing in time with a phonograph cylinder—thus partially carrying out what Muybridge had proposed a little earlier. His letter to Edison, written "as a brother scientist and inventor," described his new camera. It drew forth a formal acknowledgment—not from Edison but from one of his staff—and a request for drawings of the camera. The drawings were sent. There is no evidence that they were seen by Edison and in 1910 Edison stated in an affidavit that he had not seen the original letter.

Even had he done so, it seems unlikely that events at West Orange would have been affected, so loose were Friese-Greene's ideas when compared with the down-to-earth practical experiments of Dickson, to whom Edison had turned over the bulk of the actual development work.

By 1889 Edison was not yet clear of his electrical interests and was, in fact, preparing for their amalgamation into the Edison General Electric Company. He was already deeply involved in the huge iron-crushing project on which he was to lose a fortune in the 1890s. And he was, moreover, immersed in preparations for a visit to Britain and France.

He and his wife crossed the Atlantic in August, Edison sitting on deck by the hour and watching the waves. "It made me perfectly savage to think of all that power going to waste," he later admitted. The visit was, quite fortuitously, to have a considerable effect on the development of Edison's movie apparatus, even though it was prompted by the International Exposition in Paris where the Edison companies' display occupied a quarter of the United States exhibit. The set piece was a forty-foot high model of an incandescent lamp made up of 20,000 separate bulbs. On either side the American and French flags glowed out in colored lamps while crowded below them the space was, as one writer put it, "filled to overflowing with the varied fruitage of Edison's ripened thought." The fruitage included examples of the firms' products grouped round the giant Jumbo dynamo that had appeared in the city eight years earlier.

The display was no more than could be expected since the Edison empire had become one of the most important in Europe. More significant was the treatment given to the man himself. Made a Companion of the Legion of Honour in 1881, he was now raised to its highest rank. When he visited the Opera House—one of the first to be lit by incandescent lamps—the President of France offered his private box, and saw to it that Edison was allowed to sit in the prompt box at stage level to watch the ballet. Gustave Eiffel, whose eponymous tower, suggested by the massive framework built for the Statue of Liberty, had just been completed, invited the Edisons to his office at the top of the tower and here, more than 900 feet up, they listened to a two-hour performance by Charles Gounod. The City of Paris honored the visitor at a special dinner in the Hôtel de Ville. Louis Pasteur, then at the height of his fame, was delighted to show him round the newly opened Pasteur Institute. "Buffalo Bill," Colonel Cody, presenting a "Wild West" display on the outskirts of Paris, welcomed him to a special performance. He was honored by what appears to have been a somewhat uproarious party at the offices of *Figaro*, which had for some while been lit by his lamps. And it was only after unavailing efforts that he failed to meet Count Villiers de l'Isle-Adam, the French man of letters who had made Edison the central figure in one of the world's first works of science fiction, *L'Eve Future*.

More significant was the dinner given on 19 August to honor the fiftieth anniversary of Daguerre's revelation of his own photographic process. At the dinner Edison probably met Dr. Etienne Jules Marey, the French photographer in whose "photographic gun" moving pictures were produced by the use of a circular glass plate revolving in one direction and a circular metal disk with shutter hole revolving in another. Certainly Edison later visited Marey's shop and was shown another device in which pictures appeared in sequence one under the other. "I knew instantly that Marey had the right idea," he told Albert Smith, the movie historian, and while returning across the Atlantic he penciled out a rough sketch of the machine he wanted.

Back in West Orange, Edison was greeted by an exuberant Dickson who triumphantly demonstrated an apparatus showing Dickson himself coming forward while his voice was heard saying: "Good morning, Mr. Edison, glad to see you back. I hope you are satisfied with the kinetophonograph." Subsequent accounts are contradictory. Dickson maintained that the machine threw a picture on a screen. Edison, giving evidence on oath, was equally emphatic that "There was no screen;" yet he later wrote to a film historian: "If I made such an answer, I certainly misunderstood the question."

However, he now began to concentrate on the use of a strip of pictures, one beneath the other, such as he had seen in Marey's workshop. Development appears to have progressed with two slightly different aims. Dickson was anxious that the moving pictures should be thrown on to a large screen, while Edison, correctly sensing that something less ambitious could be perfected more easily and more quickly, concentrated on what was to become the successful "kinetoscope." This was an improved form of the peep-box on which he had issued a caveat in 1888, equipped with electric motor and a fifty-foot band of film which was pulled beneath the magnifying glass through which it was viewed. He patented the machine in the United States in 1891—although making the disastrous mistake of failing to do so elsewhere; continued with its development while at the same time trying to devise a satisfactory method of projection onto a screen; and in 1893 set up in the yard outside the laboratory the world's first movie "studio."

An oblong wooden structure, pivoted so that it could be turned to take advantage of the sun as a windmill can be turned to the prevailing wind, it was painted black inside and out. The roof of the Black Maria, as it was called, could be opened to let the sun stream in, and in lieu of the sun there were magnesium lamps and twenty arc lamps. Into this primitive studio Edison enticed all manner of performers. The dancer Carmençita came to perform for an hour. The strongman, Eugene Sandow, came to be filmed while expanding his chest from forty-seven inches to sixty-one and while holding on his chest a platform on which stood three large horses. The boxers Corbett and Courtney came to battle before the Edison camera while circus animals included performing dogs, bears, trained lions, and monkeys. The kinetoscope in which the films were shown

> attracted quite a lot of attention at the World's Fair in Chicago in 1893 [Edison later claimed] but we didn't think much of it until we found that two Englishmen who had been interested in the exhibit, finding that I had carelessly neglected to patent the apparatus abroad, had started an independent manufacture on a considerable scale.

In spite of this lapse, Edison was now quick to file further patents in the United States. On Saturday, 14 April 1894, Alfred Tate opened the first battery of kinetoscopes—ten machines—in New York. Two bright businessmen set up the Kinetoscope Company which bought the machines from Edison and began to license the use of what the Press called "The Wizard's Latest Invention." Only a few months passed before Grey and Otway Latham, two brothers who operated a licensed kinetoscope, began their own efforts to project onto a screen, which could be viewed by large numbers, the moving pictures that in the kinetoscope could be seen by only one viewer at a time.

The Lathams failed to interest Edison whose attitude is revealed by his statement, reported to Terry Ramsaye, researching for his history of the movies.

> No [he said]. If we make this screen machine that you are asking for, it will spoil everything. We are making these peep-show machines and selling a lot of them at a good profit. If we put out a screen machine there will be a use for maybe about ten of them in the whole United States. With that many screen machines

you could show pictures to everybody in the country—and then it would be done. Let's not kill the goose that lays the golden egg.

However, the Lathams went ahead. They started their own laboratory, developed their own machine, and on 21 April 1895 demonstrated their "panopticon" to journalists in New York. "If they exhibit this machine, improve on what I have done, and call it a kinetoscope, that's all right. I will be glad of whatever improvements Mr. Latham can make," Edison told the *New York Sun*. "If they carry the machine round the country, calling it by some other name, that's a fraud, and I shall prosecute whoever does it. I've applied for patents long ago."

Meanwhile, suddenly realizing the threat of the screened film, Edison moved with all his old speed and decisiveness, solving a major problem of projection by buying up the patent of Thomas Armat's cam movement device which stopped and started the filmstrip more efficiently. He was reluctant to incorporate in his own projector this work of someone else; but the race had to be won and on 23 April 1896 the first public demonstration of Edison's "vitascope"—using Armat's invention—was held in Koster & Bial's Music Hall, New York. From then onward the story of the movies became the story of rival companies leapfrogging each other with improvements and helped by the fact that in 1891 Edison had failed to patent his camera outside the United States.

Meanwhile, Edison still continued with his efforts to marry up sight and sound. One of the first demonstrations was given in the new century at the Orange Country Club to which Edison invited a number of his neighbors. They were, they were told, to witness "an improvement in the motion picture," and when a conventional drawing-room interior was thrown on the screen there was no indication that anything revolutionary was to be expected. Then a figure in evening dress walked toward the center of the screen, and raised his hands. The audience saw his lips begin to move and, as the words were formed, they heard him speaking, the sight perfectly matching the sound. Next a girl played "Annie Laurie" on the violin and a woman sang "The Last Rose of Summer." But this was only the beginning.

"The lecturer dropped a china plate on the floor," said one of the audience. "You heard not only the initial crash, but the lesser noise of the flying fragments. A bugler came on and sounded the reveille; there was the screech of a whistle; and ... some dogs were led on, and their barks were clearly heard as they scampered round the stage." Scenes from the *Chimes of Normandy* were seen and heard, as well as a politician trying to make a speech to his constituents while being prompted from behind. Verdi's "Miserere" was seen and heard and the program was completed by the "Star-Spangled Banner."

In spite of his success, Edison finally dropped "the talkies" for other things.

The problem of actual synchronization was the least difficult of my tasks [he subsequently explained]. The hardest job was to make a phonographic recorder which would be sensitive to sound a considerable distance away, and which would not show within range of the lens. You get some idea of the difficulty if I make this comparison—if you estimate the volume of sound at a distance of one foot from the recorder at one hundred you find that at a distance of two feet it diminishes to twenty-five.

But if sound linked to sight seemed impractical there were other possiblilities and Edison began to experiment with an alternative, the throwing on the screen of captions, either separately or superimposed on the picture.

We used to experiment with stopwatches on various types of mentalities, trying to strike a fair average of time to all for a given impression to register [he wrote in his diary]. We picked children and old persons, clerks, mechanics, business men, professional men, housewives, and exhibited titles with varying numbers of words. When we showed more than six or eight words at a time it was a revelation to see how many failed to get any connected thought at all.

He also investigated the educational possibilities of the film. "I had some glowing dreams about what the camera could be made to do and ought to do in teaching the world things it needed to know—teaching it in a more vivid, direct way," he wrote.

He tested his theories by practical experiment. In this case two classes of children, one of boys and one of girls, both aged under fifteen, were taught exclusively from films for a while and Edison was confident that the experiment supported his beliefs. He was once asked which was more important, eye or ear.

The eye [he replied]. Light travels quicker than sound, and the eye absorbs ideas instantly. It is my firm conviction that a large part of education in coming generations will be not by books but by moving pictures. I have tried this out in experimenting with children; and the results have been astonishing. Children don't need many books when they are shown how to do things. They can learn more by some kinds of moving pictures in five minutes than they can by the usual kind of books in five hours.

Once the early teething troubles of the new apparatus had been dealt with, Eastman and Edison had little contact with each other and not until May 1907 when Eastman visited the West Orange research laboratories did they again meet. The following year they signed an agreement which brought one a regular flow of royalties and the other a regular stream of business.

In order to carry out the arrangement at least ten different parties had to be satisfied [Eastman's biographer has pointed out]. First, Edison and Eastman, so far as the furnishing of the film and collection of royalty were concerned; then the seven proposed licensees under the Edison Company patent, so far as their relations with the Edison Company and between themselves were concerned and the relations between the Edison Company and Eastman were concerned; and then the renters, so far as their relations to the seven licensees and the Edison Company were concerned. In addition, the whole matter had to be planned so that all of the agreements would be legal, as well as what the renters were to do, and so as to remove it, as far as possible, from any attack on the ground of being an attempt to create a monopoly or in restraint of trade.

The complicated financial and patent arrangements, which marked Edison's involvement with the improved phonograph, as well as his development of movie apparatus, came toward the end of a long story which had begun, as we have seen, during the latter part of the 1880s. When it was only part told he had experienced a shock which drastically changed his attitude to research, which allowed him to continue with the phonograph and the movies, only peripherally connected as they were with electricity, but which helped to draw his main energies away from the electric light industry he had done so much to create.

The amalgamations of 1889 had provided him with ready cash and more spare time. If they had diminished his personal control at least the name of Edison remained. Then, without his knowledge, a move was made in the early 1890s by Henry Villard to amalgamate the Edison interests with those of the Thomson-Houston Company. It was his secretary, Alfred Tate, to whom details had been "leaked," who broke the news. "I have never before seen him change color," Tate wrote. "His complexion naturally was pale, a clear healthy paleness, but following my announcement it turned as white as his collar."

The new consortium was the General Electric Company. Edison's name came nowhere, and while he appreciated that control had, perhaps necessarily, passed from the inventors to the financiers, bitterness almost unbalanced him. A few weeks later, when Arthur E. Kennelly, joint discoverer with Oliver Heaviside of the ionospheric layer surrounding the earth, was still on his staff, he turned to Tate with words which revealed his feelings.

Tate, if you want to know anything about electricity, go out to the galvanometer room and ask Kennelly. He knows far more about it than I do. In fact, I've come to the conclusion that I never did know anything about it. I'm going to do something now so different and so much bigger than anything I've ever done before, people will forget that my name ever was connected with anything electrical.

Of course he knew about electricity, and he knew that he knew. But now he was to work all out on a project very different from electrical research.

Chapter Ten

Fresh Fields
to Conquer

Long before the experiments in the Black Maria began to foreshadow the motion picture industry, Edison had tentatively embarked on a new major enterprise in a field he had never before entered. From the automatic vote recorder to the idea of joining sound to moving pictures, his interests and inventions had developed along a visibly logical route. The electromagnets of the vote recorder had been exploited to introduce improvements in the stock printer and the telegraph system. From this, a growing knowledge of how electricity could be manipulated led on to the multiplex systems. The demands of the furiously competitive world of the telegraph corporations into which they led encouraged him to devise the electro-motograph, and from here it was a small step to provide the telephone, for some while an apparent threat to the telegraph, with the magic of the carbon button. The variable conductivity of materials such as carbon led to the tasimeter, its testing in Wyoming and Edison's subsequent concentration on electricity and the incandescent lamp. The phonograph had sprung directly from the inspiring noises of the automatic telegraph and in turn had led, through Muybridge, to the movie camera.

Compared with these links in a chain, the new venture to which Edison turned, the concentration of low-grade iron ore deposits, was apparently an industrial problem lacking any connection with previous activities. Yet the ore-milling project, financial disaster though it became, was a typical example of the way in which Edison so often responded to a national need. It was also more important for another, unrelated, reason. From the layout of Edison's ore-crushing mills, Henry Ford once said, there sprang his concept of mass production, the basis of American predominance in the twentieth century.

Not until the early 1890s, when the formation of General Electric hardened his resolve, did the project begin to take shape, yet Edison had been ruminating on the idea, jotting down sketches in his notebooks, and generally turning over its possibilities since the day, more than a decade previously, when he had been attracted by a patch of black sand on a Long Island beach. Curious to find out what it was, he had taken a small sample back to the laboratory. Here he had discovered that myriads of its minute black grains were attracted by a magnet— the magnet which was to bridge the gap between the Edison of the telegraph wire and the Edison of the ore-concentration plant.

By the early 1880s, the supply of iron ore to the steel industries of the Eastern seaboard was already raising problems. If it came from the comparatively small but high-grade deposits to the west, heavy transport charges were involved; that from the east, nearer to the mills, was low-grade and had to be subjected to an expensive concentration process before it could be used. Edison, aware of the problem, asked himself a simple question. Why should it not be possible to let a stream of the crushed ore fall past a magnet which would deflect the iron to one side while the rest dropped straight down. This solution—not unlike a crude version of electromagnetic separation with which physicists were to separate fissile uranium 235 from the superabundant uranium 238 in the 1940s— had been tried more than once. Until now, however, every attempt had been balked by the mechanical problems involved. But these were of the size which Edison felt born to tackle and in 1880 he lodged a patent for a magnetic ore separator, the first of nearly sixty.

The following year he set up a pilot plant on the south shore of Long Island. Its usefulness came to an end when a major storm removed the deposits in a matter of hours. A second plant was set up on Rhode Island. It produced about 1,000 tons of concentrate but this was not of usable quality. Edison's interest then waned for a few years, reviving only when the formation of Edison General Electric in 1889 gave him time and money for further large-scale efforts.

He now moved with the same care and meticulous collection of facts that had characterized his other work. First he constructed a particularly sensitive magnetic needle which would reveal the presence of even small concentrations of iron ore. Then he organized the survey of a strip of land to discover how extensive the low-grade deposits to the east really were.

Once the facts were known, he began to buy, and in the wild and wooded country of northern New Jersey there developed a new kind of mining village. It was christened Edison, an isolated outpost linked with the outside world only by a branchline which the Central Railroad of New Jersey had built from Lake Hopatcong. Homes for the workers were designed by "the Old Man" himself, lit with electricity and fed with running water. Edison himself lived in "the White House" throughout the week, driving down to Glenmont on Saturday night and returning early Monday morning. Below the 3,000 surrounding acres of his land there lay 200 million tons of low-grade iron ore; below another 16,000 acres there was a further 1,000 million tons. "These few acres alone," he claimed, "contained sufficient ore to supply the whole United States iron trade, including exports, for seventy years."

In the early 1890s his plans began to crystallize. So far, his inventions had brought him only moderate rewards. The scheme for magnetic concentration of iron ore would, he optimistically hoped, be a good deal more profitable. "In six or eight years," he told the *Scientific American* in April 1892, "I shall take out $10 or $12 million worth of ore a year, at a profit of about $3 million a year clear."

The mills themselves were a considerable technical achievement. In their design and construction he had argued that manual labor should be reduced to a minimum, that the initial job of breaking up huge multi-ton rocks as big as a house should be tackled as far as possible by machinery, and that moving conveyor belts should carry the ore through the successive stages of its treatment.

The result was a mill whose heart was a series of gigantic rollers, studded with metal protruberances and revolving at a speed of sixty miles an hour. The rollers broke up into pieces the size of a man's head the six-ton boulders fed on to them. The ore was then carried on belts to further rollers, broken down into still smaller pieces and eventually, having passed through successive stages, turned into powder which was fed past 480 magnets. Each stage took the process of iron concentration one step further, much as the cascade process was to produce enriched uranium for the Manhattan Project half a century later.

Once the iron had been concentrated it had to be processed into usable form and for the task Edison designed and built a separate factory which would turn out briquettes at the rate of one a second. He himself took an active part in everything, and an incident in the briquetting works shows Edison the man within the casual attitude of the times. Part of the machinery consisted of a lever held down by a powerful spring and a four-foot rod an inch in diameter.

> While I was experimenting with it, and standing beside it [he once recalled] a washer broke and that spring threw the rod right up to the ceiling with a blast; and it came down again within just an inch of my nose, and went clear through a two-inch plank. That was "within an inch of your life," as they say.

Characteristically, Edison loved living on the site, although he always ensured that he was back home at Glenmont for 4 July. His son Charles once wrote:

> This was the day dad really devoted to us kids. Just at daybreak on the morning of the Fourth he would come into our bedroom and wake us up. The firecrackers were ready; we never stopped to dress. Barefoot, we all dashed madly to the lawn, and there in the dawn we had a splendid time setting off the day's first firecrackers.

Father got a lot of amusement out of lighting firecrackers, throwing them at our bare feet and making us dance when they exploded. He had it all his way one Fourth. After that we ganged up and made him take off his own shoes and stockings and do his dancing on the lawn while we three lighted firecrackers at his feet.

Then it was back to the White House at Edison once more. Returning to it years later, and sitting on the porch, Edison himself reflected: "I never felt better in my life than during the five years I worked here. Hard work, nothing to divert my thoughts, clear air and simple food made my life very pleasant. We learned a great deal. It will benefit someone some time."

Unfortunately, it was of precious little benefit to Edison. Technically, the works were a triumph. The economics of operation went according to plan and he was able to supply iron briquettes, of the kind needed by the blast furnaces, at a price of about six dollars a ton. When the great scheme was planned this would have assured him of a constant flow of orders. Then, as the mills were getting into production, huge deposits of rich iron ore were discovered in the Mesabi range of north Minnesota. Investigation soon showed that the deposits were not only rich and extensive but could in some cases be mined by comparatively cheap open-cast methods. The going rate for iron ore dropped by more than a third and Edison was reluctantly forced to admit that with a selling price of less than four dollars a ton, his mills could run only at a loss.

When he was finally forced to close them down, he had poured $2 million into the project. Most of the money had come from the sale of his General Electric shares, and when the project failed the shares were worth double that sum. Reminded of the fact by his business associate, W. S. Mallory, Edison ruminated for a few minutes, pulling his right eyebrow as he often did when thinking over a problem. "Then," Mallory remembered, "his face lighted up and he said: 'Well, it's all gone, but we had a hell of a good time spending it.'"

The ability to bounce back was due partly to Edison's own abundant self-confidence, partly to the fact that he learned from mistakes and misfortunes. In designing the ore-concentration plant the magnetic basis of the process had been a strong link with past experiences; but in designing equipment to handle huge quantities of rock he had built up a solid body of knowledge in an entirely different field. Now he wondered how he could make use of it.

It was here that luck took a hand. In 1898, cement rock was discovered in New Village, forty-five miles west of West Orange, and Edison quickly bought 800 acres of the land below which it had been found. Limestone was brought in from a nearby site and by 1902 a newly constructed Edison works was turning out cement. By 1905 it had become the fifth largest in the United States.

Inevitably, Edison did more than work along accepted lines. When he entered the cement business it was accepted that the rotating kilns in which the raw materials were burned could be no more than 60 feet long. When it was revealed that the New Village mills were to be 110 feet long, the idea was ridiculed as impractical. But a few years later Edison had successfully extended his kilns to 150 feet and was using half the previous amount of coal to produce the same amount of cement. The pre-burning process of grinding he revolutionized by adaptations from his iron-ore mills, while the antiquated but normal process of moving the cement rock in fleets of carts was abandoned for the use of five-ton steam shovels.

Edison's interest was practical. He was already convinced that whether the future lay with the electric car or the gasoline-driven vehicle, the day of the horse was past. America, he believed, would soon need something better than the dirt roads which had served earlier generations. Concrete might be the answer.

That was not all. It would be ingenuous to claim that any of his multifarious activities were driven on mainly by social motives. Yet Edison's understanding of the ordinary people was real enough. He had kept his humanity, and what would today be called his social conscience, even while moving through the labyrinthine world of the robber barons, and he felt deeply about what he saw as the evils of the time. One of those evils was the slum tenement area of New York—and that of other great cities of which he had less personal knowledge. Cement, in Edison's vision, was the ideal material for building small, cheap but clean buildings in which families could live without the disadvantage of tenement life.

Neither of his two ideas was particularly successful. The first one-mile stretch of concrete road, laid near New Village, lasted only a year, and its successors had little better luck. Later, the trouble was found to have been caused largely by the clay base of the road. A few miles of concrete road laid elsewhere survived into the 1950s.

More visionary was his ambition to build concrete houses, a project for which he filed a patent in August 1908.

> The object of my invention is to construct a building of a cement mixture by a single moulding operation [this said]—all its parts, including the sides, roofs, partitions, bath tubs, floors etc., being formed of an integral mass of a cement mixture. This invention is applicable to buildings of any sort, but I contemplate its use particularly for the construction of dwellings, in which the stairs, mantels, ornamental ceilings and other interior decorations and fixtures may all be formed in the same molding operation and integral with the house itself.

The mold was to be made of cast-iron sections bolted together to form a hollow into which the concrete was poured. Strength was given at strategic points by wires embedded in the material, one of the first examples in the United States of reinforced concrete. When the concrete had set, the sections were to be unbolted and taken away to reveal the complete house. To ensure that the heavier ingredients did not sink to the bottom and leave the lighter solidifying above, a jelly-like colloid was added.

Edison announced what he was going to do rather than what he had done, thus laying himself open to attack by *Punch*, the London journal whose job was poking fun at the frailties of the world. While Edison's announcement had paralyzed the building trade it had stimulated activity elsewhere, *Punch* rejoiced.

> The more extravagant party in the London County Council talk of laying liquid cement mains in Suburban London. It would be a great boon, they argue, to the ratepayer to be able to turn on the cement, just as nowadays he turns on the water for the garden hose. If unexpected guests come for whom there is no room in the house, if a fowl house or dog kennel should be required, if the householder has ambitions towards a billiard room, if a porch or conservatory, or even a summer-house, should need to be built, if the roof begins to leak in a storm or (as in some cases it has done) becomes restless, if the garden wall must be raised to keep next door from staring—in fifty different emergencies a ratepayer would find an ever-ready supply of cement most useful. All he would have to do would be to send down to the local ironmonger for the moulds, stick them up, and then leave the tap running into them, with perhaps the youngest boy to keep an eye on it.
>
> We would like to suggest that the cement tap ought to be coloured red, so that it be not confused with the water tap. Cement, however liquid, is not a good thing to water the garden with or to boil the potatoes in.

There was a good deal in similar vein, much of it due to garbled reports. Edison replied, rather pained, to *Concrete and Constructional Engineering*, reiterating his plans and adding:

> I have not gone into this with the idea of making money from it, and will be glad to license reputable parties to make molds and erect houses without any payments on account of patents, the only restriction being that the designs of the houses be satisfactory to me, and that they shall use good material.

Shortly afterward he attended in New Jersey the construction of the first concrete house, and watched as the huge mechanical mixer was secured in place. Concrete was carried up by bucket conveyor to a reservoir at roof level and poured into the molds which were filled within six hours. Six days later they were removed and it was then only necessary to put in doors and windows, and to install and connect the plumbing and lighting fixtures.

Edison followed this early experiment with a more elaborate plan, for houses built incongruously "in the style of Francis I, richly decorated with designs that

185

would be prohibitive because of their cost were they in stone." No inside decoration would be necessary while "if it is desired to heighten the inside effect, tinting can be resorted to ... the roof imitates tiling and can be painted to suit the owner's taste." But in spite of concrete's alleged insulating qualities, claimed to reduce fuel costs by three-quarters, the house failed to become popular.

Edison's move into more prosaic industry during the 1890s and the first years of the twentieth century tended to harden the ambivalence felt about him in the scientific world. It had initially arisen not so much because he was a technologist rather than a scientist, a distinction which divided gentlemen from players well into the twentieth century and in places still separates engineers from the rest. More important were his views on the relative virtues of the sciences and the arts in education, and on the values of bookwork and handwork, views that sometimes lurched over from commonsense practicality into philistinism. There was also at times the feeling that he had not kept his hands clean enough in dealing with the "robber barons" whom he openly despised but to whom he had been willing to sell his expertise. Above all there was one maddening fact: he had on occasion, and notably about the subdivision of electricity, been right while the qualified scientists had been wrong. Some, like Preece, were big enough to admit it. Others were smaller.

The mixed feelings that Edison's record produced were heightened when the electrician went in for ore milling and the telegraph engineer went into housing. The situation was remarked upon as early as 1893 by Hermann Lemp, a scientist who had worked with Edison and who attended the splendid banquet which closed the Chicago Exposition. While it was in progress, he wrote in his memoirs,

> I noticed my old chief, Thomas Edison, coming in with Samuel Insull and seating himself at an out-of-the-way table. He was not a delegate to the Congress and so did not occupy a place among the speakers. The truth was that the theoretical physicists, particularly those from Europe, looked down upon Edison as not belonging to them, and in many instances acted snobbishly towards him. Though there was no slight intended on this night: they knew of Mr. Edison's dislike of public appearances and had left him to himself. So it did my heart good to see Professor Helmholtz, one of the greatest luminaries among physicists, step down from the speaker's table and come over and shake hands with Edison.

Helmholtz's action had been noted and during a lull in the speeches later on, the diners began to rap the tables and call out: "Edison, Edison, Speech!" Edison, who always hated speaking in public, merely got up, smiled, bowed, and sat down. The calls were renewed. Once more, the response was the same.

The feeling that Edison was moving completely into the industrial world was by no means justified. He still experimented on a number of imaginative if smaller enterprises. His notebooks continued to be filled with ideas and his head with ambitious projects.

In 1893, when his hopes of the ore-concentration project were still high, he had far-reaching plans for the time when it would be running itself.

> I shall [then] turn my attention to one of the greatest problems that I have ever thought of solving. That is, the direct control of the energy which is stored up in coal, so that it may be employed without waste and at a very small margin of cost.

Pointing out that only 10 percent of the energy in coal was converted into power in a boiler, he forecast something better.

> It would enable an ocean steamship of 20,000 tons to cross the ocean faster than any of the crack vessels now do, and require the burning of only 250 tons of coal instead of 3,000.... I have thought of this problem very much, and I have already my theory of the experiments, or some of them, which may be deemed necessary to develop this direct use of all the power that is stored in coal.

How would it be done? All he would say was that

> the coal would be put into a receptacle, the agencies then applied which develop its energy and save it all, and through this energy electric power of any degree desired could be furnished.

Nothing came of this particular vision—nor of a curious scheme which arose from his consideration of the great mass of iron ore near the Ogden mills. Solar flares, he knew, produced disturbances in the earth's magnetism which were recorded on magnetometers in the world's observatories. Why, he asked himself, should it not be possible to increase this effect enormously "by utilizing a vein of magnetic iron ore and running round the pot of ore several miles of wire, forming an induction circuit into which pulsating electric currents would be thrown by any disappearance of the earth's magnetism." He had therefore erected telegraph poles on either side of the ore hill and linked them by many miles of wire.

He was still awash with ingenious ideas, and some had been worked out in detail. One, which he patented, exploited magnetism for the unusual task of transmitting power at high velocity without the noise of toothed gearing or the limitations of normal pulleys and belting. His plan was to magnetize both pulleys and belts: the pulleys by suitable windings connected to an electric current and the beltings either by the same method or by studding them with iron bars which closed the circuit at the pulleys. The result was "a magneto-mechanical device capable of replacing toothed wheels in nearly all cases of power transmission."

Although virtually nothing came of this particular idea—he always pointed out that of even the most potentially useful inventions only a small percentage was ever used—it was an indication of his irrepressible eagerness to exploit for practical use the facts of science that the nineteenth century was still revealing. When in 1895 Roentgen announced his discovery of X-rays, Edison at once plunged in and lost no time in producing them himself. With his usual systematic methods, he set about investigating their possible uses and four men were put to work making various chemical compounds, amassing about 8,000 crystals of different substances, and checking that several hundred of them would fluoresce under the X-ray.

It was to Edison that Michael Pupin, the Yugoslav-American physicist, turned a few weeks after Roentgen's announcement. Pupin had been taking radiographs and to him there came a well-known lawyer whose hand had been peppered with shot-gun pellets. The physicist was unable to locate the pellets with the materials available. A calcium tungstate screen from Edison altered that; the pellets were located within a few seconds and the hand was saved.

> I also found that this tungstate could be put into a vacuum chamber of glass and fused to the inner walls of the chamber [Edison said] and if the X-ray electrodes were let into the glass chamber and a proper vacuum was attained, you could get a fluorescent lamp of several candlepower. I started to make a number of these lamps, but I soon found that the X-ray had affected poisonously my assistant, Mr. Dally, so that his hair came out and his flesh commenced to ulcerate. I then concluded it would not do, and that it would not be a very popular kind of light: so I dropped it.

But it was too late to stop Clarence Dally from becoming one of the first X-ray martyrs. First his hands, then his arms, were amputated in an effort to prevent the disease from spreading. He died a few years later.

Edison exhibited the fluoroscope at the New York Electrical Exhibition in 1896, the first time that the American public was shown the vital new tool of future medical research. But for Edison's personal intervention the demonstrations might have been postponed or canceled as the large induction coil in the exhibit was found to be damaged when it was being installed on a Saturday afternoon. The exhibition was to open on Monday morning, all the machine shops would be closed until then, and several miles of wire on the coil had to be entirely rewound. Mallory later related:

> Edison would not consider a postponement, so there was nothing to do but to go to work and wind it by hand. We managed to find a lathe but there was no

power; so each of us, including Edison, took turns revolving the lathe by pulling on the belt, while the other two attended to the winding of the wire. We worked continuously all through that Saturday night and all day Sunday until evening, when we finished the job.

Four sets of apparatus were exhibited and so arranged that, as the *Electrical Engineer* put it, visitors to the exhibition would "be able to inspect their own anatomy." Between 3,000 and 4,000 had by the end of the exhibition

had a momentary glimpse of their own bony structure—for the first and last time in this life. Many of them flinched when they got in front of the screen and refused to look, either at their own bones or at anybody else's. Some crossed themselves devoutly after a fearsome glance, although, as a matter of fact, the great majority came out all smiles and laughter.... The crowd was in reality delighted to find the show neither horrid nor awful but tickling and mirthmaking.

In the semidarkness of the demonstration tent only a few recognized the operator who from time to time took over the controls. It was Thomas Edison.

The display gave the Edison-baiters their opportunity. Although it might be laudable for Roentgen to make a scientific discovery, applying it could be a different matter, as London's *Pall Mall Gazette* made clear. Its editor wrote:

It is now said, we hope untruly, that Mr. Edison has discovered a substance— tungstate of calcium is its repulsive name—which is potential,whatever that means, to the said X-ray, the consequence of which appears to be that you can see other people's bones with the naked eye and also see through eight inches of solid wood. On the revolting indecency of this there is no need to dwell, but what we seriously put before the attention of the Government is that the moment tungstate of calcium comes into anything like general use it will call for legislative restraint of the severest kind.

Perhaps the best thing may be for all civilized nations to combine to burn all works on the Roentgen Ray, to execute all the discoverers and to corner all the tungstate in the world and whelm it in the middle of the ocean. Let the fish contemplate each other's bones if they like.

However, if the British made sophisticated jokes about X-rays, others were more serious. Surely, it was suggested in New Jersey, the use of X-rays in opera glasses at theaters should be banned.

Edison was also closely following another scientific development as significant as the discovery of X-rays. This was application of the electromagnetic waves he had failed to recognize in 1875, the "etheric force" which Hertz had produced in Germany and which many men, notably Enrico Marconi, were conscripting for the wire-less telegraphy of the future. Edison seems to have felt no regret that

189

events in this field had passed him by. Even he had to recognize that there were only twenty-four hours in each day to work and think. First the phonograph, then the incandescent lamp, then more purely industrial activities had claimed his concentration. He was wise enough to realize, intuitively, that an attack on the mysteries of telegraphy without wires would demand resources in time and money which he would have difficulty in finding. He had the incandescent lamp and the phongraph; some bright young man like Marconi could have the wireless.

In December 1901, there came news that the first messages had been sent across the Atlantic from southwest England to Newfoundland. It was followed by Edison's laconic comment: "If Marconi says it's true, it's true." He was unable to attend the celebratory dinner held for Marconi by the American Institute of Electrical Engineers a month later, but telegraphed an apology. "I would like to meet that young man who has had the monumental audacity to attempt and succeed in jumping an electric wave across the Atlantic," he went on. The toastmaster, T. C. Martin, having read the telegram from Edison, said that he had been talking to Edison who

> had thought that some time there might be daily signals across the Atlantic without wires, but that he did not know when, and being preoccupied he did not think he would have time to do it himself. He said to me, "Martin, I'm glad he did it. That fellow's work put him in my class. It's a good thing we caught him young."

Edison did more than send good wishes. Two years later he sold to Marconi— allegedly for "a song"—the patent on wire-less telegraphy he had filed on 23 May 1885. This was of crucial importance since it buttressed Marconi's position in the litigation—as extensive as it had been on the telegraph and the incandescent lamp—which followed the birth of radio.

Edison's interests were now being drawn off into a different field once again. As with the telephone and the incandescent lamp, he saw that an age was ending, in this case the age of horse-drawn transport. Just as he had believed that on the railways electricity would oust steam, a belief which took longer than he estimated before it was justified—so he now saw the passing of the horse, at least in the expanding cities of America. His enthusiasm for a powered car was rooted, as were so many of his ideas, in the belief that it must necessarily add to the quality of life—a belief to which he might today add a few qualifications. There was, he wrote, no reason why horses should be allowed in cities at all.

> They are not needed [he argued]. The cow and the pig have gone, and the horse is still more undesirable. A higher public ideal of health and cleanliness is working toward such banishment very swiftly; and then we shall have decent streets, instead of stables made out of strips of cobblestones bordered by sidewalks. The worst use of money is to make a fine thoroughfare, and then turn it over to horses....

Whether they would be replaced by the electric car or by the gasoline engine was still in doubt. The electric vehicle suffered from the heavy weight of the battery that powered it, and the fact that the largest practicable battery had to be recharged after a limited number of miles. However, to a man of Edison's unqualified optimism, these were merely problems to be solved by hard work and common sense. For long, the electric car was in his eyes the ultimate answer, and as late as 1914 he and Henry Ford were announcing joint plans for production.

Edison might have pushed on more determinedly with his plans had it not been for Henry Ford, a man who from the closing years of the nineteenth century was convincing even such electricity enthusiasts as Edison that whatever might happen to their proposed vehicle, a gasoline car would be running on the roads beside it.

Ford had joined the Detroit Edison Illuminating Company in 1891 and by 1896 had become its chief operating engineer. He had already built his first "horseless carriage" and was at work on his second although he had so far received few words of encouragement from anyone. In August he came, with the chief managers and engineers of the Edison companies, to the Edison Convention at the Manhattan Beach Hotel. After dinner the talk flowed over into general business affairs, and the prospects of the electric car. Someone mentioned that Ford had made a gas car and Ford was asked to explain. Edison did not hear very much because of his deafness, so Ford moved next to him.

There are various versions of what happened next. Ford has given more than one: so has Edison. But though details differ, the general course of their first encounter is clear. So is the impact that each man made on the other. Edison, in his fiftieth year, and despite his undiminished high spirits "old" in a way that twentieth-century man is not old at fifty, was on the lookout for bright young men of imagination. Ford, with a weakness for success and a predilection for hero-worship, found the Messiah he was looking for; possibly because, as he once put it, Edison's electric light had lengthened man's day, so that he could increase his production and his producing efficiency. Even at the age of thirty-three, Ford had no need to hitch his wagon to a star; had it been necessary, he would have chosen Edison.

Now, meeting for the first time, the younger man was bombarded with questions. How many cylinders did his car have? Was the gas exploded by contact or by a spark? Ford later recalled:

> He asked me no end of details, and I sketched everything for him for I have always found that I could convey an idea quicker by sketching than by just describing it. When I had finished, he brought his fist down on the table with a bang, and said: "Young man, that's the thing; you have it. Keep at it. Electric cars must keep near to power stations. The storage battery is too heavy. Steam cars won't do either

for they have to have a boiler and fire. Your car is self-contained—carries its own power plant—no fire, no boiler, no smoke and no steam. You have the thing. Keep at it."

That bang on the table, Ford said, was worth worlds to him. "No man up to that time had given me any encouragement."

Ford, like Edison, would no doubt have found his own way forward whatever the difficulties. But as his biographer notes, "he was now listening to the most renowned inventor in the world, who endorsed his little car with unmistakable emphasis. Edison's verdict confirmed his faith and sent him into action."

Edison's friendship with Ford, who not long afterward left the Detroit Edison Illuminating Company to found his own empire, endured until Edison's death more than a third of a century later. It was close, to the extent that each admired the other; also, despite their differing views on the electric car, their ambitions did not conflict. But it was hardly close in the accepted manner of friends. Affable as he was to employees and to business colleagues, Edison could afford little time for friendship, even had he felt the need for it, which is unlikely. It was not only growing deafness which tended to make him walk alone.

However, both men believed in the coming age of the automobile, however it was to be powered, and they were soon discussing a small but more efficient battery which would power the starter and generator of a gasoline-driven vehicle as well as the vehicle's other electrical equipment.

As I began to explain to him what I wanted, I reached for a sheet of paper and so did he [Ford later recalled]. In an instant we found ourselves talking with drawings instead of with words. We both noticed it at the same moment and began to laugh. Edison said: "We both work the same way."

Shortly afterward he was thinking of something more ambitious, a storage battery which would radically depart from the pattern followed for the previous forty years. "Beach," he said, turning to a colleague, "I don't think Nature would be so unkind as to withhold the secret of a *good* storage battery if a real earnest hunt for it is made. I'm going to hunt."

As early as 1861 Glaston Planté had immersed lead plates in an acid electrolyte to produce what was called "stored electricity." Some years later Camille Faure had made a pasted-plate battery and in 1881 Charles Brush added various improvements which turned the battery into a practical proposition for electrical vehicles. It was possible, with Brush's battery, to get eight watts of electricity for every pound of material used, and the main aim of subsequent developers and inventors was to increase the electricity per pound, since weight was so important. Other aims were to reduce manufacturing cost and increase the life.

Man into Myth

Edison at work in the chemical department of the West Orange laboratory about 1890

Top: *Edison and George Eastman with one of the early Kodak cameras*

Above: *Edison operating a motion picture machine—possibly one of his rival's—in the library at West Orange, 1897*

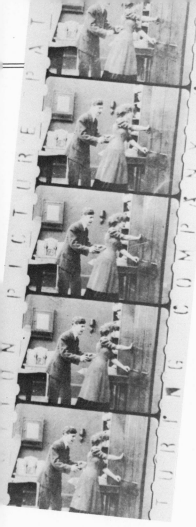

The Kinetoscope,
showing the eight reels
of film and, on top of the
machine, the viewing
aperture

ove: *Two Edison film strips:
t on the left was made for the
Kinetoscope in 1891,
that on the right for the
Cinematograph in 1911.
picture frames on both were an
ich wide and three-quarters
of an inch deep, the only
erence between the two being
that the Kinetoscope film
had to be printed more
densely*

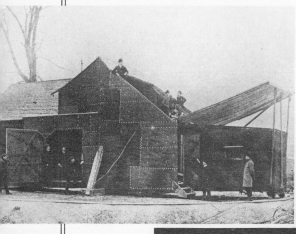

Left: *The world's first film studio, the "Black Maria" erected in the grounds of Edison's West Orange laboratory. Part of the construction's roof could be opened and the whole building rotated in order to catch the sunlight*

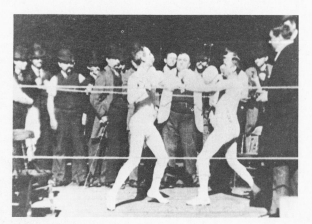

Three films made after Edison had developed the Kinetoscope: a boxing bout photographed inside the "Black Maria," 1895; Amy Muller dancing in the "Black Maria," 1896; "Going to the Fire," shot in the streets of Newark, 1896

196

A line of kinetoscopes in the etoscope, Phonograph & Graphophone Arcade, San Francisco, 1894

Below: *Edison's iron-ore quarry with stock house on right, at the New Jersey and Pennsylvania Concentrating Works, Edison, New Jersey*

Left: *Edison with George Meister— the assistant secretary of Edison's laboratory—in an electric Studebaker, 1909*

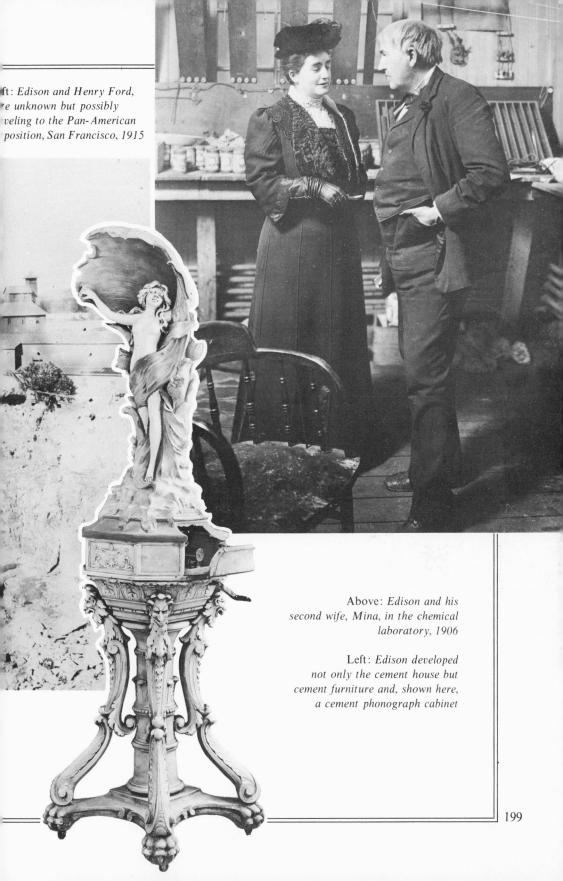

*ft: Edison and Henry Ford,
e unknown but possibly
veling to the Pan-American
position, San Francisco, 1915*

Above: *Edison and his
second wife, Mina, in the chemical
laboratory, 1906*

Left: *Edison developed
not only the cement house but
cement furniture and, shown here,
a cement phonograph cabinet*

199

Top: *Edison on board U.S.S. New York, 1916 with (center) Josephus Daniels, Secretary of the Navy and (right) Admiral Henry T. Mayo. Standing in background are members of the Naval Consulting Board*

Above: *Edison and other members of the Naval Consulting Board on the Citizen's March through New York, 13 May 1916*

Above: *Singing the "Star-Spangled Banner" in front of Edison's rebuilt laboratory, Flag Day, 1915*

The wreckage of Edison's
West Orange factory,
severely damaged by fire
on 14 December 1914

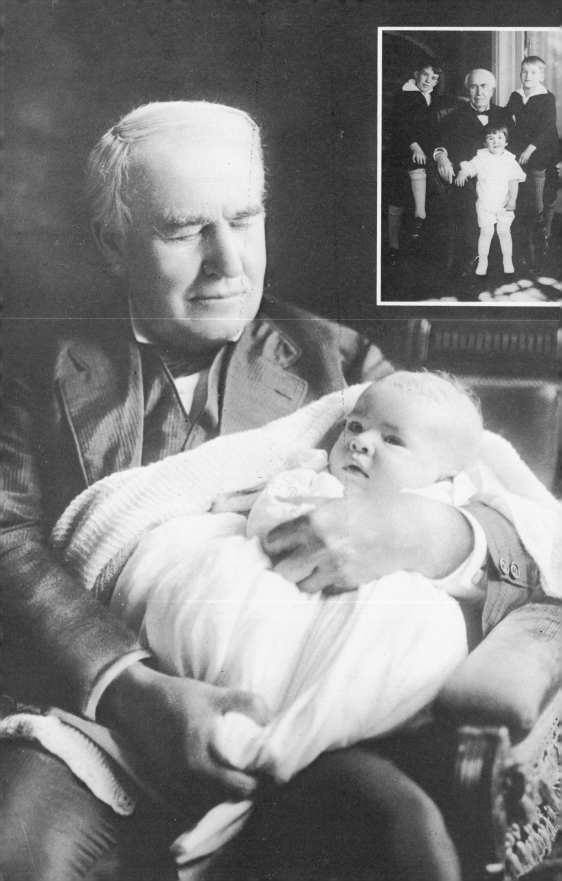

Inset: *Edison with his daughter Madeleine's three sons—Thomas, John and Peter Edison Sloane, 1926*

Left: *Edison and his first grandchild, Thomas Edison Sloane, 1916. "He had assured us that the baby wouldn't move," says Mrs. Sloane. "He didn't"*

Right: *Edison with his wife at the unveiling of a tablet at Menlo Park which marks the spot where his laboratory stood*

Overleaf:
Edison sleeps during a camping trip 23 July 1921. Seated in background, Harvey S. Firestone (left) and President Harding (right)

Top left: *Edison and Harvey S. Firestone with car and supply truck, about to set out for their camping trip to West Virginia, 28 August 1916*

Bottom right: *Edison and his party stop to examine an old grist mill in the West Virginia mountains, 21 August 1918. (Left to right) Edison; Harvey Firestone, Jr.; John Burroughs; Henry Ford; Harvey S. Firestone; (seated) R. J. H. de Loach*

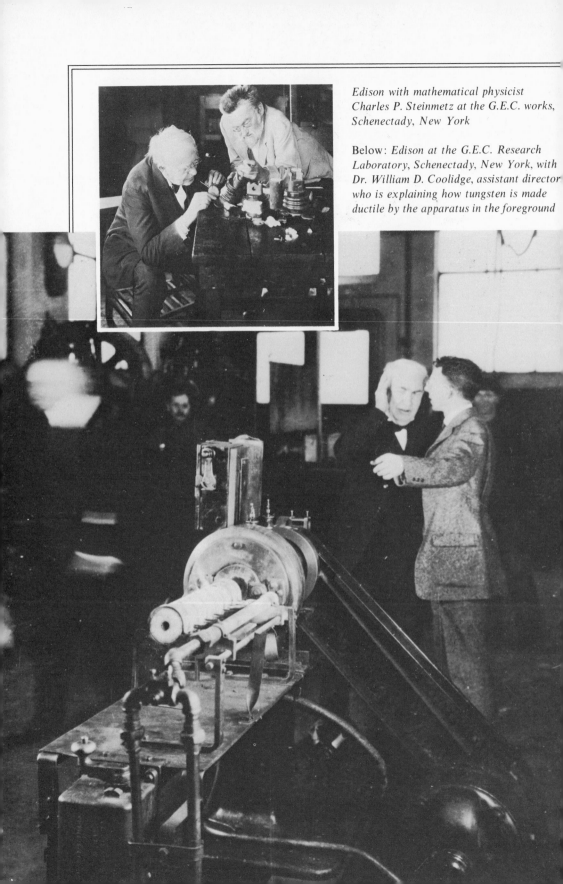

Edison with mathematical physicist Charles P. Steinmetz at the G.E.C. works, Schenectady, New York

Below: *Edison at the G.E.C. Research Laboratory, Schenectady, New York, with Dr. William D. Coolidge, assistant director who is explaining how tungsten is made ductile by the apparatus in the foreground*

...son with Irving Langmuir
...amining a vacuum tube in
...research laboratory,
...henectady Works. On left,
...orge Morrison, manager of
...General Electric Lamp
...rks in Harrison, New
...sey

...ow: Group at Dearborn,
...Dctober 1929, showing (left to
...ht) Henry Ford, next
...he locomotive engineer;
...s. Edison; Edison;
...sident Hoover; Mrs. Hoover
...d Mrs. Ford

Edison writing in his notebook, 1928

The conventional lead storage battery is charged by passing a current through it, the result being that oxygen is added to the positive plate and taken away from the negative plate. When the battery is used, the positive plate gives up its oxygen, the negative plate is oxidized, and an electric current is generated.

Edison believed that it was unsatisfactory for contact between a metal and an acid to be the mainspring of a battery, although a technically feasible alternative would not be easy to find. But once he had decided on the "earnest hunt" he set to work with his normal energy.

> About 7:00 or 7:30 A.M. he would go down to the laboratory and experiment, only stopping for a short time at noon to eat a lunch sent down from the house [W. S. Mallory has said]. About six o'clock the carriage would call to take him to dinner, from which he would return by 7:30 or eight o'clock to resume work. The carriage came again at midnight to take him home, but frequently had to wait until two or three o'clock, and sometimes return without him, as he had decided to continue all night.

This went on seven days a week and was still the regular practice when Mallory visited him after five months.

> I found him at a bench about three feet wide and twelve to fifteen feet long, on which there were hundreds of little test cells that had been made up by his corps of chemists and experimenters. He was seated at this bench testing, figuring and planning. I then learned that he had thus made over 9,000 experiments in trying to devise this new type of storage battery, but had not produced a single thing that promised to solve the question.

Eventually, he began to get what he was looking for, and by the summer of 1902 was planning a 5,000-mile reliability trial of a vehicle powered by a battery which would take a vehicle 100 miles on a charge. After the test, of which Edison gave a hair-raising description—"when we turned a sharp corner I thought every time the machine was going over"—he was confident that the electric machine was the vehicle of the future. He intended to build "a good machine," which would run at a steady twenty-five miles per hour, and he had no fear of faster gasoline-driven vehicles.

> If they go faster than my machine, I will be able to go downhill as fast as they dare to [he said] and for hill climbing the electric motor is just the thing, so I will beat them there. On rough roads they will not dare to go any faster than I will; and when it comes to sandy places, I am going to put in a gear of four to one which I can throw in under such circumstances, and which will give me 120 horse power of torque, and I will go right through that sand and leave them way behind.

Although plans for the vehicle got little further, his new batteries were put on the market in 1904. It was soon evident that something was wrong. Many batteries developed important weaknesses in service, and although the company decided to replace them Edison himself halted production.

At this point most men would have thrown in their hand. But not Edison. Once more, he started from scratch, investigating the weakness in the battery by the laborious trial-and-error methods that had always served him so well. He worked the same irregular hours, followed the same irregular habits that at times drove even his more devoted workers to the edge of rebellion but only to the edge not over it.

Rosanoff, who had solved the problem of a better wax for phonograph records, has recalled one typical occasion on which Edison had phoned some of his staff, told them that he would be working that evening, and suggested that they join him.

> Sometime between midnight and one o'clock Charlie Edison complained to his father that he was getting "kind of dopey" and would like to take a *little* nap. "Well," said the Old Man, "if you *got* to sleep, go lie down under the table in the corner; nobody will step on you there." Charlie carried out the suggestion literally and was soon fast asleep on the floor under a table. About two in the morning, Mrs. Edison drove over, worried about Charlie. The Old Man emphasized that Charlie was safe where he would not be stepped on. Charlie's sleeping on the hard floor did not meet with Mrs. Edison's approval. She next disapproved of Mr. Edison's expectorating on the floor and politely offered to provide a spittoon, but he declined, saying that the floor itself was the surest spittoon because you never missed it. Charlie, however, was taken home to sleep.

Cobalt was among the scores of materials investigated in such unconventional working conditions, and after three prospecting geologists had found supplies near Charlotte, North Carolina, Edison decided to inspect for himself. He did so by driving to Charlotte from Orange with his son Charles and three assistants in two White steam cars, and camping *en route* where no inns were available. Like the eclipse expedition nearly three decades earlier, it was a jaunt justified by work.

The technical problems solved before the improved battery was eventually put on the market were numerous and intricate, none more so than the production of nickel flakes only one twenty-five-thousandth of an inch thick. Edison had decided that this wafer thinness was essential. Making them satisfactorily was a triumph not only of chemistry—a sphere in which Edison is rarely given enough credit—but of technological expertise. According to Byron Vanderbilt in his account of Edison as a chemist:

Ten revolving copper cylinders were carried by a crane, dipped alternately into copper- and nickel-plating baths, and sprayed with water after each plating. This process of alternate plating was repeated 125 times in a five-hour period. Later the process involved the laying down of 150 layers of each metal. The 0.0075-inch thick copper-nickel composite consisting of 250 layers was then stripped from the copper cylinders and cut into $\frac{1}{16}$-inch squares. The copper was then dissolved chemically, first with aqueous ammonia containing a mild oxidant but later with a solution of sulfuric acid saturated with nickel sulfate. The thin flakes were washed, centrifuged, and dried over steam coils. This flake nickel was so thin it would float in air like thistledown. A bushel weighed only four and a half pounds despite the fact that nickel is 8.9 times heavier than water.

The end product of which these nickel flakes formed an essential part was a battery which gave 100 miles to the charge compared with the 50 miles for most lead batteries. It could not be so easily damaged by overcharging or by standing unused for weeks on end, while its life was many times that of the lead battery. For a while it almost seemed as if the electric car would be a major competitor with gasoline and as late as January 1914 Henry Ford announced, after a three-hour business talk with Edison at Glenmont: "We're getting ready to put an electric car on the market." Six months later Edison was telling the *Wall Street Journal* that he was still working with Ford on an electric car which he estimated would cost between $500 and $700. "I believe that the electric automobile will be the family carriage of the future," he added. "All trucking must come to electricity." But his friend Ford, as much as any man, eventually tipped the scales in favor of gasoline.

The automobile, electric or gasoline, was not the only revolutionary method of transport for which Edison saw a great future during the first decade of the century. As far back as 1885, according to a statement he made thirty-eight years later, Gordon Bennett had given him $1,000 to investigate the possibilities of flight.

> I constructed a helicopter but I couldn't get it light enough. I used stock-ticker paper made into gun cotton and fed the paper into the cylinder of the engine and exploded it with a spark. I got good results, but I burned one of my men pretty badly and burned some of my own hair off and didn't get much further.

Four years later, talking of the experiments to a Paris journalist he spoke of the "great discouragement" he had met.

> You can make up your mind however, that these fellows who are fooling around with gasbags are wasting their time. The thing can't be done on those lines. You've got to have a machine heavier than air and then find something to lift it with. That's the trouble, though, to find the "something." I may find it one of those days.

Understandably enough, he was quick to see the significance of the Wright brothers' success, and was confident about what was to follow. "Within five years," he forecast in 1908, "air ships"—a word which he used for aircraft rather than dirigibles—"will be carrying passengers across the oceans in eighteen hours, by which time aerial flights will have been commercialized. The North Pole can, and will, be reached in a forty-hour trip." Less than a year later he was telling the *New York Times* that in ten years "flying machines will be used to carry mails. They will carry passengers, too, and they will go at a speed of 100 miles an hour. There is no doubt of this."

Yet some of his detailed prophecies were a combination of lucky guesses and Jules Verne ideas. Thus to the *New York Times* he said:

> If I were to build a flying machine I would plan to sustain it by means of a number of rapidly revolving inclined planes, the effect of which would be to raise the machine by compressing the air between the planes and the earth. Such a machine would rise from the ground like a bird does. Then I would drive the machine ahead with a propeller.

Not too wild a guess at the contemporary vertical-lift aircraft. So, too, he believed in what he called the "helicoptal" plane, although from his description it is a little difficult to consider exactly what he had in mind. "With a series of aeroplanes forming a circular machine with each plane running off on a tangent from the central motor, and two small—very small—propelling planes (the smaller the greater velocity) and the gyroscope principle introduced as a steadying factor, the problem is solved."

Chapter Eleven

War Against
the U-Boats

Edison was into his early sixties when his new-type storage battery was finally put on a sound commercial basis. Its development during the previous few years had brought forth his last major group of patents, but he was by no means finished. A man with his ebullience and everlasting optimism would go on to the end, and he had in fact two more roles to play, two more exciting chapters to live through: of technological adviser in the war he hoped would never happen, and of legendary symbol from an America which was already part of history.

In 1914 he was still working hard—not only at West Orange but at the Florida home at Fort Myers, which he visited each winter and onto which he had built a replica of the New Jersey laboratory. Not quite as hard as the young telegraphist of Civil War days, but still doing enough for two men of his age and still razor sharp. The reason, he always maintained, was simple enough: no more sleep than he needed, and no more food. He had never required much sleep and he believed that most other people slept too long. "Everything which decreases the sum total of man's sleep increases the sum total of man's capabilities," he argued. "There really is no reason why men should go to bed at all, and the man of the future will spend far less time in bed than the man of the present does, just as the man of the present spends far less time in bed than the man of the past did." On food, he was less wild, although his distrust of all uncooked foods on the ground that they might carry typhus was something of an obsession. On quantity he was adamant, proudly claiming that he ate only "five ounces to a meal, three times a day, including the water in the food," admitting that a man doing physical labor might require twice as much, but still certain that Americans could, on average, cut down their food intake by two-thirds. "They do the work of 3-horse-power engines and consume the fuel which should operate 50-horse-power engines," he explained. A supporter of prohibition when it was introduced in 1919, he took alcohol sparingly, and was known to turn away even the most persuasive old employee when he learned that charity would be spent on the bottle.

There was also his deafness, which he still maintained was an asset to him.

> On Broadway I can be as undisturbed as the average man can be in the deepest recesses of the most silent forest. That has been, and is, an advantage to me. It saves me from many interruptions and much nerve strain. My nerves are quite steady. I have known people to say, "Edison hasn't any nerves." I have just as many and as delicate nerves as anybody else, but they are not constantly disturbed by noises.

He needed to be alert since he was to feel the impact of the European war, which broke out in August 1914, sooner than most Americans. His industries depended on supplies of raw materials from Britain, who now kept them for her own use, or from Germany whose exports were stopped by the British blockade. However, before he was seriously to feel the effects of the war he experienced a setback that could have been disastrous for a man in his late sixties.

At 5:30 on the evening of 9 December a big explosion shook the film finishing house at the West Orange factory. The cause was never discovered but within a few minutes the entire floor was ablaze, the flames being fed by chemicals and highly inflammable rolls of film. Before the first firemen arrived not only the finishing house but the neighboring buildings were on fire and all effort had to be concentrated on saving the most valuable areas. Edison, hearing the explosion from his house, had immediately driven down to the site and was soon directing the firemen. At the same time he was seen jotting down notes in one of his small pocketbooks. They were his plans for rebuilding the factory.

Even disaster could not suppress the irrepressible Edison and as the blaze increased he called to his son Charles: "Where's your mother? Get her over here, and her friends, too. They'll never see a fire like this again." And after he had retrieved from the wreckage a framed portrait of himself, its glass cracked by heat, its frame charred by flames but the portrait itself unmarked, he wrote on the mount: "Never touched me."

Mrs. Edison had been helping to rescue from her husband's office dozens of what seemed at first sight to be comparatively valueless chairs, desks, pictures and statuary. "I knew," she explained later, "that there was apparatus in it on which he had been working for years. If anything would discourage him, the loss of this building would do it." As it happened, things were not as bad as that. The office was eventually saved, although the firemen had to fight for hours to get control of the blaze, hampered by lack of water and by the cold weather and flurries of snow. Roasting on one side and freezing on the other, was how some of them described it.

It was midnight before the fire was in hand and by then ten buildings, roughly three-quarters of the complete works, had been destroyed. Much equipment was saved, as well as $40,000 worth of phonographs. But Edison was unable to rescue a number of glass jars in the phonograph building. Each contained several thousand dollars worth of diamonds used for recording.

His notes on rebuilding, taken when the fire was still blazing, were no mere gesture. "Although I am over sixty-seven years old, I'll start all over again tomorrow," he told reporters the following morning as he surveyed the wreckage— a reaction that brought a congratulatory letter from President Wilson. The production of films did in fact restart the following day and by the early spring the entire factory had been rebuilt: but only after Edison had forced from the local authorities the promise to build a new 50-million-gallon reservoir. It had been a mistake not to ensure that fires could be properly fought, and Edison was still ready to learn from mistakes.

Even more significant was another outcome. He had observed how handicapped the firemen were by their inability to see through the dense smoke. One night a few months later people living in the valley to the east of Glenmont began telephoning the police to ask about an astonishing light coming from the house. Investigators found Edison and his family playing with a portable 3-million-candlepower searchlight he had invented to help fire fighters. It would burn for two hours, was powered by a battery weighing only two pounds and could throw its effective beam several miles.

By the time that the rebuilt factories were getting back to work he was being seriously affected by the war in Europe where, the German advance into France having been halted, both sides were facing the prospect of a long campaign of attrition. Although reluctant to admit that the United States could or should be involved, Edison seems to have known within his marrow that American industry would be crucially affected and that men like himself would be called upon to exercise all their technological expertise.

He himself was better equipped than most to meet the challenge. Two years previously he had visited Germany and had returned to the United States immensely impressed with what he had seen there. One point in particular had struck him. This was the way in which the Germans used scientific research to ensure the maximum utilization of their raw materials.

> At the great Badesch chemical works, 200 research workers are engaged upon investigations that mean continual new commercial products. The thorough, careful way in which this concern makes researches in therapeutics may well be considered by every American manufacturer, whether he makes synthetic indigo, sulfonal, food products, or metal goods. In this country we go in for the obvious products, the ones we can get quickly and easily. Just as we throw away slabs at the lumber

mill, so we lose by-products simply because we do not think it worthwhile to bother with intricate processes or undiscovered uses. The Germans have proved that it is worthwhile, that it pays to study details.

This was of course the basis on which Edison had built up his own successes. But one thing was slightly ironic. Those who admitted the German superiority in technological industry usually saw it as springing from the pride of place Germans gave to scientific education. Edison, one of the few American leaders who had built up industries on German-style foundations, had enjoyed precious little education, scientific or otherwise.

It was of great benefit to America that during the early years of the European war, as public opinion gradually veered from neutral to pro-Allied, a man of Edison's eminence should continue to hammer home the lesson of German industrial efficiency without thought of political implications. But there was a price to be paid for this straight-from-the-shoulder attitude and the criticism of home methods whoever the criticism might offend. It came, in fact it probably only could come, from a man who was politically simplistic. The Edison who could upbraid American industry for not being as efficient as the Germans, had little interest in, and less understanding of the underlying causes which bring about the rise and fall of political parties and of nations, and was therefore vulnerable to the more specious ideas and ideologies which crises bring forth. It was perhaps inevitable that during the upheaval of World War I he should occasionally make a fool of himself.

As early as October 1914, little more than two months after the German armies had invaded France and Belgium, he was being reported by a Dearborn journal as stating that one cause of the war had been the commercial rise of Germany, that the Jews had been largely responsible for German business success and that "the militarists which govern the country do their bidding." Protests were immediate and loud. He tried to explain himself, saying that he had merely wanted to praise the Jews for their ability. "However," say Henry Ford's biographers in dealing with the incident, "a number of Edison's letters to Ford and E. G. Liebold [managing editor of the *Dearborn Independent*] show a distinct anti-Semitic bias."

There is other evidence. Edison once said that his definition of a successful invention was "something that is so practical that a Polish Jew will buy it" while to a post-war employee he once claimed: "There are lots of bad Jews. Soon as they knew I was going to use a special wax for my cylinder records they hacked up the price on me. But I fooled them—I changed over to celluloid."

Not too much should be made of this. More important in the *Dearborn* articles of 1914 is the ignorance of conditions in Germany which Edison's statements betray. Anti-Semitic feeling in Germany and Austria-Hungary was often present even before the coming of Hitler. Yet interviewing Edison on a subject of which he knew little was only a beginning. During the 1920s, as he became among the best-known of all Americans, he was regularly interviewed on almost every subject under the sun, a forerunner of the contemporary T.V. pundit.

The first effect of the war on the Edison enterprises concerned carbolic acid or phenol. This was essential to the manufacture of phonograph records and no less than a ton and a half of it was being used every day in the West Orange factories. Until the late summer of 1914 the acid, in the form of crystals packed in metal drums, had been imported from Britain and Germany and a 100,000-pound shipment was ready to sail from England when the embargo was imposed. The chemicals were taken off the ship and Edison was left with supplies for only ten weeks and no hope of more to come.

The accepted belief was that although carbolic acid was made from coal it was impossible to make it economically, if at all, from the coal that was mined in the United States. In any case, while it could easily be imported there seemed little inducement to develop for commercial production any of the methods that had been tried out in the laboratory.

Edison acted; partly, of course, because he had no intention of letting his record-making factories grind to a halt, but also because he was piqued by the experts' persistent statements: that it would probably be impossible to make the acid in America at all, that if the job could be done it would be uneconomic, and that there would, in any case, be many months of delay before any new factory could start production.

For three days Edison studied in detail the half dozen known methods of making synthetic carbolic acid. He narrowed the choice to two, personally experimented in his laboratory with both, and finally chose the sulfonic acid process. Next he consulted the manufacturing chemists. Would they build a factory and provide him with supplies? And, if so, when could he expect deliveries? The chemists were cautious. There were no precedents for such a factory and Edison insisted that it must be right first time; there was no room for mistakes. One firm thought that he might have supplies within six months. Seven, eight, and nine months were the more usual estimates.

Now came the Edison touch. First he called together forty men, chemists and draftsmen, and told them what he wanted. He divided them into three groups which were to work eight-hour shifts. He himself was to live in the laboratory. As the plans progressed they were discussed and altered, tested bit by bit where possible. Edison worked round the clock, snatching the occasional hour or two's sleep on the top of an old roll-top desk barely two feet wide and not long enough for a full-length stretch. "This was the only bed 'the Old Man' knew for weeks," said John F. O'Hagan, who was working with him. "He arose one night after about two hours' sleep and in the darkness quickly put on his hat, only to jump with consternation and to throw his hat against the wall as he felt some creeping creatures on his head. While he slept a family of mice had made their home in it."

A week after the men had been set to work the plans were complete. Next a site was chosen at Silver Lake, not far from Orange, and building work began without delay. On the eighteenth day after the plans had been authorized the new Edison factory produced 700 pounds of carbolic acid. Output quickly increased and before long was in excess of their needs. The surplus was sold to other factories, but success brought its own problems. Edison was soon, reported the journal *Weekly Drug Markets*, "besieged by brokers and agents, eager to buy up all, or any part, of his output." While some wanted phenol for genuinely pharmaceutical purposes, this was not the whole of the story. Phenol could be converted into picric acid, the essential element in a number of explosives.

Phenol was only the first shortage. Benzol was a close second. "I made application to a large steel plant, offering to put up a benzol plant myself, but they would not listen to me although I offered to rent a small section of their property, make the whole thing myself and pay them eighteen cents per gallon," Edison said later. "They would incur no expense and I would be paying them for what they were throwing away, but their board of directors would not permit me to do this." Eventually Edison was making his own benzol, just as he was soon to be making other wartime essentials.

It was obvious that he was capable of overcoming wartime shortages as efficiently as he had overcome the problems of distributing electric light, and American industry was quick to seek his help. The rubber companies needed anilines and within a few months Edison was making the material from benzol. "Then there came an appeal from the fur dyers," he wrote. "They could not get a pound of what is called paratphenylene-diamine. They wanted to know if I could make that for them. I put up a plant and furnished all the dye for them. Then they wanted aniline salt, so I put up a plant for that as well."

In thus producing what America had always imported, Edison was helping to break the near monopoly of Germany in dyes. He was indiscreet enough to say so and was immediately attacked by Germans in the United States for "un-neutral" actions. In fact, few men could have been more neutral minded than Edison. One of his basic beliefs was that an America armed in the right way would be able to pursue her own isolationist course without fear of attack, and even after the *Lusitania* had been torpedoed by the Germans in May 1915 with the loss of 1,198 lives, about 100 of them American, he still believed that "going into the war is the very last thing we should think of at present." Grasping

218

at a straw, he suggested that the sinking, which had profoundly shocked America, might have been the act of an isolated submarine commander operating on his own initiative.

> But even admitting that the destruction of the *Lusitania* was by Governmental order [he argued], there lies in the dreadful circumstances, so far as I can see, no good excuse for making us go to war. I cannot imagine anything which Germany could do which would make that an advisable or admirable course.... How could we help by going into the war? We haven't any troops, we haven't any ammunition, we are an unorganized mob. I cannot believe that Germany even seriously fears our entrance.

His reasoning brought the pained reply from a correspondent in the *New York Times* who retaliated that "even so logical a thinker as Mr. Edison gets muddled when he joins the peace-at-any-price crowd." However, Edison persisted in his support of purely defensive forces. Since the summer of 1910 he had in fact been at work adapting his famous storage battery for use in submarines. In July four young submarine officers had visited his laboratory and described the danger of chlorine poisoning, then a major hazard faced by submarine crews, and for the next two years Edison worked on the development of a battery for underwater use. Finally, he discovered a method of substituting potash solution for sulfuric acid which removed the danger of chlorine poisoning and, as an unexpected bonus, increased the range of a submarine from 100 to 150 nautical miles. Two U.S. vessels were being equipped with the new Edison batteries when the European war broke out and there is an exhilarating account of Edison visiting the Brooklyn Navy Yard where the battery was being automatically rolled up and down to simulate the effect of a submarine's movements. "Make her rock faster," he exhorted the men testing it. "Give her a big tip. Bump her. Do anything you want to her. I've tried everything and you can't faze her." What is more, he said, the battery would last "four years, eight years—it will outlast the submarine itself."

While Edison was anxious to give first priority to the United States, he appears to have adopted to potential buyers the attitude exemplified by his dealings with Western Union and Jay Gould. When the King of Siam visited Orange after the fire of December 1914 he brought his Military Attaché and both inspected the new submarine battery. Other countries were also interested. "Three times a year for four years, a representative of the Krupp people, acting for the German Government, came to the U.S. to see what progress was being made with the submarine battery, having found the lead battery not entirely satisfactory," said his manager in May 1915. "Mr. Edison, however, would not sell a set of the cells until he was satisfied they were perfect and would stand no improvement. This battery was not perfected until last September, or a month after the war started."

It was not the first time that Edison had considered the application of science to war. Some years previously, when it had seemed possible that the United States might be embroiled with Chile, he had expounded on the potential of electricity to "make gunpowder and dynamite go sit in humble obscurity with the obsolete flint arrowhead and call him brother." Every electrician, he said, would have his own plan for making the enemy "electrically uncomfortable."

His own—when he let his imagination run loose—was for the defense of individual forts. He told the *New York World:*

> In each I would put an alternating machine of 20,000 volts capacity. One wire would be grounded. A man would govern a stream of water of about 400 pounds pressure to the square inch, with which the 20,000 volts alternating current could be connected. The man would simply move this stream of water back and forth with his hand, playing on the enemy as they advanced and mowing them down with absolute precision. Every man touched by the water would complete the circuit, get the force of the alternating current, and never know what had happened to him. The men trying to take a fort by assault, though they might come by tens of thousands against a handful, would be cut to the ground without any hope of escape. Foreign soldiers undertaking to whip America could walk round any such fort as mine, but they could never go through it. It would not be necessary to deal out absolute death unless the operator felt like it. He could modify the current gently, so as simply to stun everybody, then walk outside his fort, pick up the stunned generals and others worth keeping for ransom or exchange, make prisoners of the others if convenient, or if not convenient turn on the full force of the current, play the hose once more, and send them to the happy hunting grounds for good.

The picture, declared the *Scientific American* on reprinting the interview, was "a most beautiful and attractive one."

More seriously, he had in the early 1890s cooperated in the invention of the Sims–Edison torpedo, an electrically controlled weapon which could be maneuvered two miles ahead of a battleship by means of wires. "It is a very pretty and destructive toy," he once said, "but it is not in that kind of thing that I take pride."

His pacifism quickly came to the surface but it was as quickly overcome by patriotism. Thus during the Spanish-American War of 1898 he made what appears to have been his first contact with the U.S. Navy Department. They should, he proposed, fill shells with a mixture of calcium carbide and calcium phosphate. When these hit the water near enemy ships they would explode, ignite, then burn for several minutes, thus making enemy ships visible up to five miles away.

He had other ideas. Even before the Wright brothers had fulfilled their appointment with destiny on the sandhills of Kittyhawk, Edison had a vision of what manned flight, as well as electricity, could mean for the wars of the future.

Torpedo boats will be despatched two miles ahead of a man-of-war and kept at that distance under absolute control, ready to blow up anything within reach. I believe, too, that aerial torpedo boats will fly over the enemy's ships and drop a hundred tons of dynamite down on them. A $5 million warship can be destroyed instantly by one of these torpedoes.

I can also conceive of dynamite guns [he went on]. I have no intention of ever devising machines for annihilation, but I know what can be done with them. Nitro-glycerin is one of the most dangerous substances that man can deal with. Touch a drop of it with a hammer and you will blow yourself into the hereafter. Iodide of nitrogen is even more dangerous. While experimenting with explosives in magnetic mining, I made some of them so sensitive that they would go off if shouted at. Place a drop on the table and yell at it and it will explode.

Even without such ideas, Edison was obviously a man to question as the sinking of the *Lusitania*, despite President Wilson's pronouncement that America was "too proud to fight," inevitably brought the prospect of war nearer. In May he was interviewed at length by a *New York Times* journalist and on the 30th the newspaper devoted three pages to his ideas of what the country should do. He did not believe in large standing armies, nor even a large standing navy, but he did believe that the United States should build large numbers of ships, commission them, and then put them into dock under a skeleton maintenance force until they were needed.

I advocate not only the construction of an enormous number of submarines [he went on] ... to be held in readiness for operations, not to be kept in commission, but our manufacture at once of a vast supply of harbor defense mines and the construction of many vessels properly equipped to plant them hurriedly in case of emergency.

Then came a proposal that was to have important repercussions.

I believe that in addition to this the Government should maintain a great research laboratory, jointly under military and naval and civilian control. In this could be developed the continually increasing possibilities of great guns, the minutiae of new explosives, and the technique of military and naval progression, without any vast expense.

When the time came, if it ever did, we could take advantage of the knowledge gained through this research work and quickly manufacture in large quantities the very latest and most efficient instruments of warfare.

Among those who read the Edison interview was Josephus Daniels, Secretary of the Navy, and on 7 July Daniels wrote to Edison stating that the most important needs of the Navy were machinery and facilities for utilizing the natural inventive genius of Americans to meet the new conditions of warfare. He intended to establish a department of invention and development to which all ideas and suggestions from service of civilian inventors could be referred.

> I feel that our chances of getting the public interested and back of this project will be enormously increased if we can have, at the start, some man whose inventive genius is recognized by the whole world to assist us in consultation from time to time on matters of sufficient importance to bring to his attention. You are recognized by all of us as the man above all others who can turn dreams into realities and who has at his command, in addition to his own wonderful mind, the finest facilities in the world for such work.

Would Edison act as head of the proposed Board?

Edison, Daniels said later, "replied with what the Navy would call a cheerful 'Aye, aye, sir,' saying he would volunteer for any services his country desired him to render." A few days later Daniels traveled out to Glenmont, spent two hours discussing the situation, and shortly afterward announced that Edison had agreed to become President of a Naval Consulting Board.

One of Edison's first recommendations to the Navy Secretary was that the presidents of the eleven largest engineering societies in the United States should each be asked to nominate two members to the Board. Daniels agreed and Edison was supremely confident of what the results of his proposal would be. Through Miller Reese Hutchinson, his personal representative, he announced: "You may rest assured that when this board is completed it will, in my opinion, far outrank any body of scientists and experts that was ever collected together in any single organization."

It was three months before the organization was complete and the Board gathered in Washington for its first meeting. Edison's presidency had certainly stirred the public imagination, but his most important work for the Navy was in fact to be done off his own bat, in his own laboratory. One reason was that he excelled as the individual man of ideas rather than as the committee chairman. Another was his deafness. Hutchinson, a former telegraph operator, tapped out the proceedings in Morse on Edison's wrist. Nevertheless, deafness was still a handicap in the cut and thrust of argument and in the early stages of the Board's deliberations there was a good deal of argument, much of it centered on the site of the proposed Government laboratory. A majority favored Annapolis: Edison favored Sandy Hook. His preference was not adopted and although the laboratory was eventually set up he appears never to have visited it.

Contact with the officers of the Navy Board awakened Edison rather suddenly to what modern war could mean, and soon after the first meeting he uttered a warning. "The soldier of the future will not be a sabre-bearing, bloodthirsty savage. He will be a machinist. The war of the future, that is, if the United States engages in it, will be a war in which machines, not soldiers, fight." Shortly afterward, he was hammering home the implications: "Science is going to make war a terrible thing—too terrible to contemplate. Pretty soon we can be mowing men down by the thousands, or even millions, almost by pressing a button. The slaughter will be so terrible that the machinery itself will virtually have to do the fighting." When it came to the horrors of military science already used in Europe, gas and the flame thrower, he was equivocal in a way that made little sense even half a century ago. Asked what he thought of such weapons, he replied: "They are perfectly proper for use in defense, but not for offense. A man has a right to claw, scratch, bite or kick in defending himself, but when he is on the offensive, no."

The remarks, typical of many decent men when first brought face to face with the truth of what war means, were made as Edison was going to, and returning from, the Pan-American Exposition held in San Francisco during the autumn of 1915.

A highlight of the exposition was Edison Day when, on 21 October, the thirty-sixth anniversary of the incandescent lamp was honored, the culminating event being a splendid banquet at which, remarkably for those days, everything was cooked by electricity. Tied in with the celebrations was a demonstration in the West Orange laboratory of a new phonograph into which Anna Case of the Metropolitan had already sung an aria from Charpentier's *Louise*. The recording was relayed by telephone to San Francisco, and so was the voice of Miss Case singing the same aria, live, in West Orange. Real voice, it was claimed, was indistinguishable from recorded voice.

In San Francisco Edison was also guest of honor at a telegraphers' banquet. Speeches were made through telegraph sounders, wires were stretched from table to table and at each there was a sounder connected to the general circuit, while at Edison's table a special resonator had been wired in. It was, he said afterward, the first banquet at which he was able to hear all that was said by the speakers.

Henry Ford also attended the exposition and with Edison decided to visit the nearby Santa Rosa nurseries of Luther Burbank, the famous horticulturalist. Ford wanted to discover if Burbank's new garden peas were similar enough in size to be gathered by machinery. Edison had no particular interest in horticulture; but after signing the visitor's book in which he entered "Everything" beside the query "Interests," he began to enjoy the visit, occasionally leaning toward his wife if he had missed one of Burbank's explanations and asking: "What did he say, mother?"

Burbank wrote of the visit:

> They wanted to know everything about the flowers and the plans and the program, and Edison was particularly quick to see beauty and catch the vision of what was being done and attempted. Henry Ford was just as enthusiastic, but he saw a different angle of the gardens. He wanted to know what was being done to increase production and develop new possibilities in plants; he is as keen as mustard and has the longest view into the future of any man I have encountered out of the business world of my time. The ladies said we acted like three schoolboys but we didn't care. We were having a boss time!

It was a typical Edison jaunt, an unexpected time off which he thoroughly enjoyed since it savored of a naughty boy playing truant, and he was in the right mood to accept the proposal now made by Harvey S. Firestone, the tire manufacturer who had also been attending the exposition. Edison and Ford had both been traveling in their special rail cars. Why not abandon them, Firestone urged, so that all three could travel south by automobile to San Diego where Edison had to attend another Edison Day celebration? That suited Edison exactly—"he likes to ride in a motor—as long as he can ride in the front seat. He has no use for any other part of a motor car!"

A good time was had by all, and before the party broke up Edison suggested that the three of them should take a break the following year and go camping together. No one seems to have been particularly keen on the idea. Firestone and Ford were nevertheless persuaded, the latter possibly being influenced when Edison said that the party would be completed by John Burroughs, the veteran nature writer whom Ford had always admired. Eventually Ford had to drop out for business reasons, and it was Firestone, with his young son, Harvey junior, several helpers, a good cook, and a truck carrying refrigerator and food, who set out for West Orange.

> They met the commander of the expedition [says Firestone's biographer, Alfred Lief] in front of his laboratory, wearing unpressed dark blue clothes (in contrast to Firestone's dapper, light-gray summer suit) and a flat, black hat. All the others wore caps. Edison had a truck ready with tents, floorboards, folding tables and chairs, as well as storage batteries, lamps and wiring. He called the Firestones into his Simplex car and ordered the cavalcade on. They were going to pick up eighty-year-old Burroughs in the Catskills and persuade him to see some real mountains.

Thus began a 1,000-mile camping tour which took them from New Jersey through the Adirondacks, almost into Canada and back through Vermont.

Edison supplied the equipment, including four tents, and was specially proud of a storage battery which lit up not only the campsite but each of the four tents.

A taste of what was to come was introduced on the first night

> We made our first camp by a creek [said Firestone] and no sooner had we settled than the farmer came along to know what we were doing there and to say that, whatever we were doing, we had better get off. I told him that Mr. Edison was one of the party, but that did not mean anything to him. He "allowed" that he never let any tramps or gypsies stay around on his land, even if one of them was named Edison. We settled it by giving him five dollars which was what he was after.

Edison was the undisputed leader of the party, which made a number of similar trips during the next few years, working out the route in advance, deciding on campsites, and laying down rules for the journey, one of which was that no one should shave. During the 1916 trip it was decided that the night was too cold for Burroughs to sleep out and he was driven to a nearby hotel. Seeing the hotel bedroom and bath, Firestone found the temptation too much and spent the night there.

> I did not know what was going to happen when I reached camp in the morning with Mr. Burroughs and found Mr. Edison and Harvey [Junior] alone at breakfast [Firestone later said]. Mr. Edison saw at once that I had shaved. He did not so much mind having me out all night, but he did not like that shave—the breaking of the rules. "You're a tenderfoot," he scoffed. "Soon you'll be dressed as a dude."

Edison's friendship with Ford, to be given considerable publicity through later camping trips together, was never quite as strong as some supposed. Yet it had been strong enough to bring Edison within an ace of accompanying Ford's ill-fated Peace Ship to Europe in 1915, a mission whose aim it was to urge a compromise agreement between the warring powers. Edison went with his wife to see the Peace Ship leave. Preparing to return ashore, he was told by Ford: "You must stay on board, you must stay on board." Then, according to William Bullitt, a reporter from the *Philadelphia Ledger*, Ford said with a quizzical smile, but with what the reporter described as intense seriousness: "I'll give you a million dollars if you'll come." "Because of his deafness," say Ford's biographers, "Edison couldn't hear; Ford repeated the offer but the inventor smiled and shook his head. However, he assured his friend that he was heart and soul with him."

While Edison, like Ford, long clung to the belief that peace could be arranged if only a few men of goodwill got down to the job, it did not inhibit his work for the Naval Consulting Board. He devoted an increasing amount of time to it during 1916, and in January 1917, three months before America entered the war, handed over all other work to deputies and colleagues. For the next two and a half years he worked solely to help win the war at sea.

One of his first actions after America's declaration of war was to telephone President Hibben of Princeton University and ask for four physicists to help with one particular problem. One was Karl Compton.

> Immediately upon meeting Mr. Edison and barely taking time to say "how do you do," he took out his pencil [Compton has written] and began to describe a problem which had been put to him by the Naval Consulting Board—the problem of increasing the efficiency of the driving mechanism of a torpedo so that a larger amount of explosive could be stored in it without changing its range or size. He gave me a very brief history of the development of the present torpedo, told me the conditions which an improved torpedo would have to satisfy, and told me to come back to see him when I had a solution.

Compton reported that he had found three possible fuels. Edison disposed of them in three sentences. "Fuel A can only be obtained in Germany. Fuel B has been tried but discarded because of the danger of explosions. Fuel C, which includes wood alcohol, is no good because the sailors drink the d— stuff."

Eventually Compton proposed another alternative. Edison looked at the technical details, muttered a bit, then said:

> When I don't understand work like this I get two men to work at it independently. If they agree, maybe it is all right; if they don't agree, I get a third man. Go up into room 2—and see whether you agree with a young fellow from Columbia University whom I put to work on the same problem.

Much of Edison's work for the navy was development of torpedo-detection methods. Throughout 1917 he investigated underwater telephone devices, resonators, and a variety of towed equipment, and by the end of the year, it was becoming possible to detect torpedoes at 5,000 yards.

Submarine detection was also a constant preoccupation, and here Edison's ingenuity in tackling difficult technical problems often came to the fore. Compton has given one classic example. The navy had even before the outbreak of the European war been searching for a more sensitive microphone for the work. The carbon granules normally used had too high a resistance and Edison wanted to experiment with metal granules instead. However, they were too sluggish and he therefore worked out an ingenious scheme for producing lighter ones. First he obtained a supply of hog's bristles. These he plated with a variety of metals, some by the electrolytic process used in making phonograph records, others

by cathode-sputtering in a vacuum, and yet others by covering with the con-
densation of evaporated metals. The plated hog bristles were then cut into pieces
a hundredth of an inch long and immersed in a bath of caustic potash—"the
stuff men dissolved their murdered wives in" according to Edison—which dissolved away the bristle and left only the tiny rings of metal. These minute
granules then replaced the carbon particles in experimental microphones.

Another tactic in the antisubmarine war rested on the fact that submarine
torpedoes are fired not at an enemy ship but at the place where she will be when
the torpedo intersects her line of progress. "With this for a premise," Edison
said, "I calculated and experimented and devised a method by which a ship,
no matter what her size and speed, can change her course to one at right-angles
with it, in from two-thirds to three-fourths of her length." Ships thus equipped
also had a listening device and when the sound of a torpedo being fired was
heard the course-changing device was operated.

"Like most efficient devices, it is simple," Edison explained after the war, "—
nothing more than huge hawsers, from which are swung very large conical sea
anchors. This apparatus is connected to the bow, not the stern and when
dropped overboard, simply stops the bow while allowing the stern to swing
round. The change of course is accomplished in a few hundred feet of water."

By the time the course-changing apparatus had been tested the submarine
menace had been virtually beaten—a victory partly due to a little-known, un-
spectacular, but vitally important piece of work by Edison. This was nothing
less than an early piece of operational research. Edison obtained from the navy
full accounts of each sinking by submarine since the start of the war. He then
abstracted certain details from them and plotted the results by days, by hours
of the day, by ship lanes, by ports, and by lighthouses. One startling discovery
emerged: most ships were still following prewar routes and it was on these that
most sinkings occurred. Another point was that while only 6 percent of sinkings
took place at night most vessels were still traversing the danger zone in daylight.
From the figures alone it was comparatively simple to deduce some definite
rules showing how, when, and where a ship was least likely to be sunk.

However, Edison was never a man to be satisfied with mere deductions if
experimental proof could be sought, and he therefore prepared a pegged board
covering a chart of the channels and coasts of England, Ireland, and Scotland.

This chart [says the official account of his work] was laid off in squares of forty
miles each, which is the approximate visibility of smoke from a cargo boat as seen
from a submarine in the center of the square. Each square was provided with a
peg and a pegboard. One person had the problem of taking into British or French
ports, say, thirty vessels. His opponent had a similar pegged board with thirteen
pegs representing submarines. The first player routed his ships to the various ports

at various hours, while his opponent placed his submarines at points where he thought most likely a vessel would come into his visible area, which was then considered sunk. It was found that by following certain methods these thirty vessels could be brought into port with a surprisingly small number having been seen by the submarine.

Edison's ideas for winning the sea war ranged from the almost impractical to those which had never been tried because they were so obvious. At one extreme there was the plan for scores of buoys, manned by teams of three, to be moored between 50 and 100 miles from America's east coast, and supplied with food and water for four weeks, during which time they would presumably radio submarine sightings back to the mainland. By comparison there was the information presented to the Cunard Steamship Company which conclusively showed that the use of anthracite instead of coal reduced the area within which a ship might be seen by a submarine from a circle forty miles across to a circle twenty miles across. Furthermore, removal or camouflage of masts and funnels reduced this still further to a circle twelve miles across.

Many other problems were tackled. Naval observers at the top of high masts were exposed to gases from the ship's smokestacks so Edison devised a special mask and tested it himself in a closed room filled with burning sulfur vapors. He designed nets for catching torpedoes and a more efficient periscope: better sailing lights for convoys; and he even worked out a plan for mining Zeebrugge harbor—the objective of a famous Royal Navy raid in 1918—with flat-bottomed, power-driven unmanned boats to be steered by gyroscopic rudders loaded with explosive.

Virtually all this work was carried out by Edison himself, either in his own laboratories, or in U.S. naval yards, and with the official help of navy personnel. At times, it seems, the help was grudging. Thus even the official record reports that when he was about to perfect an antisubmarine listening device, the experimental ship was withdrawn. And of his work on improved sailing lights for convoys it says without comment: "An electrician from one of the United States submarines had been detailed to assist Mr. Edison in these experiments, but while the work on the perfected model was in progress he was withdrawn and was not thereafter returned to help in the completion of the device."

These were no doubt minor pinpricks. But from the records there does come a suggestion of surprise that Edison, a man who would fail to recognize a naval tradition if he met it in broad daylight, could help the United States Navy win the war. Edison himself had no doubt about the position. "I made about forty-five inventions during the war, all perfectly good ones," he said a few years later, "and they pigeonholed every one of them. The naval officer resents any interference by civilians. Those fellows are a close corporation."

It probably mattered less than Edison would have been willing to admit. The *New York Times* had been correct when it had remarked that his appointment would "stir the public imagination." That was very largely the object of the exercise. Yet it was not only the imagination of the public that was affected. To some foreign visitors, for whom the image of the freebooting employee of the robber barons died hard, Edison was something of a surprise. Professor Fabry of Paris, head of an Allied scientific mission visiting Washington, spoke of him as "simple, direct, intelligent, unspoiled—a very much greater man than I expected to find in view of the way his name has been exploited and the kind of influences with which he has been surrounded."

And now, perhaps somewhat late in the day, the purest of pure scientists were beginning to see that beneath the jovial combination of engineer and technologist, ready to improvise at the drop of a hat to check an idea, there existed a mind anxious to sound the profounder depths of science. "He was then, at the age of seventy and more," said Robert Millikan, who in 1915 had determined the charge on a single electron,

> reading some of the new books that were then appearing in the field of pure science, and asking intelligent questions about them too. His ears were gone, but there had been no crystallizing of his mind such as occurs with some of us before we are born; with others, especially with so-called men of action, before we are forty; and with most of us, even with those who have learned to combine the art of knowing with the power of doing, by the time we are seventy.

Chapter Twelve

Man into Myth

On 24 January 1918, a handful of Edison's colleagues founded the Edison Pioneers, a group at first restricted to those who had worked with him at Menlo Park but subsequently augmented by "honorary" members who had joined him later. The act was symbolic. While the Pioneers' main activity was to hold an annual reunion on Edison's birthday, they met with the same air of resolute dedication as the officers of Washington's staff after the end of the American Revolution. Like Nelson's captains and Henry V's band of brothers, they had together survived a great experience and as disciples they epitomized the increasing veneration with which Edison was to be held throughout America, and indeed in some respects throughout the world, for the final years of his life.

By now he had not only the record of true greatness but all the personal characteristics which in Henry Wallace's century of the common man gave that greatness continuous news value. Edison's shaggy head was as universally recognized as Einstein's leonine mane a decade and more later. His voice, carrying into old age the unaffected provincial accents of his youth, spoke with a friendly familiarity. His aphorisms, thrown off genuinely enough but with inspired timing, were those which ordinary men and women understood, while his emphasis on the homely virtues of hard work, temperance, and simplicity, touched a sympathetic chord from coast to coast. No folk hero could wish for more, and it was inevitable that during the years which followed the war Edison should become the most famous of all Americans.

Fame had its drawbacks. However much he might enjoy backing into the headlines he had reluctantly to admit that the public figure has no private life, a state of affairs which eventually brought to an end the camping holidays with Ford and Firestone. During the last year of the war they had snatched a fortnight for a tour down through the Smoky Mountains into West Virginia, as far south as Asheville, North Carolina, and then back through the Shenandoah Valley to Maryland. In 1919, with the war over, they hoped for something more ambitious. They hoped also—or, more accurately, Edison hoped and the others may have—that they would recapture some of the back-to-the-brushwoods spirit of their first adventure. In fact there was not the slightest chance of this. If Ford was already famous Edison was almost a myth. The odd man out might ask, possibly perversely, "who's Edison." More typical was the small girl, asked by Edison in an isolated village if she knew who he was. "Mr. Phonograph," she replied.

Even if journalists and photographers were not automatically alerted to follow the famous trio, the trio itself did little to ward off publicity. A professional photographer accompanied them in 1918, while in 1919 the simple camping party had developed into a three-car convoy which included "the kitchen cabinet car" incorporating a large gasoline stove, special food compartments, and an icebox. By 1921, when the party set out again, simplicity was abandoned. Wives were added to the party; so were sons and daughters-in-law, while Firestone invited his old friend President Harding, with his wife, to spend a night at their campsite. The President came on a Sunday, with a secretary, six secret service men, nine movie cameramen, and ten Washington correspondents. On his suggestion, a bishop and his wife were also invited, the result being a camp of twenty tents and a fleet of trucks to deal with the commissariat. The comments of John Burroughs, who had died the previous year, would certainly have been interesting.

There was one more trip, in 1924, then they were abandoned. Firestone complained that the party had become a traveling circus and as Charles Sorenson, one of Ford's aides, has put it, "With squads of newswriters and platoons of cameramen to report and film the posed nature studies of the four eminent campers, these well-equipped excursions into readily accessible solitudes were as private and secluded as a Hollywood opening, and Ford appreciated the publicity."

Whatever Ford appreciated, Edison would have preferred the solitude. Each winter he found it for a few months at Fort Myers, but for the rest of the year he remained in the thick of the industrial fray at West Orange. Now in his seventies, he was persuaded to cut down his "working day" to sixteen hours. Although the day was interspersed by irregular cat naps, although the picture of the irrepressible Edison still busying himself as he had done half a century earlier, is more myth than reality, he did still keep a firm hand on the phonograph and other industries at West Orange. At times an interviewer would have the temerity to raise the question of retirement. "The day before the funeral" was one reply. Another indication of Edison's attitude was his answer: "When the doctor brings in the oxygen cylinder."

He was, moreover, still an innovator. An example was the Edison Question-naire, an examination paper to be completed by applicants for jobs which aroused praise and condemnation in almost equal measure. There were 150 questions which appeared to have only one thing in common: a total absence of anything connected with the job. That fact, added to the apparent irrelevance of many questions, attracted such biting criticism that Edison was forced to defend himself. The result was a series of interviews with the editor of the *Scientific American* and a long article in which he explained that finding a good executive by trial and error was too expensive.

"So I made up my mind that we should have to have a formal test of some sort," he explained. "This brought up the question of what we should look for; what is the most important qualification for an executive?" His answer was "a good memory." Of course, he went on "it does not follow that a man with a fine memory is necessarily a fine executive. He might have a wonderful memory and be an awful chump in the bargain. But if he has the memory he has the first qualification and nothing else matters."

Edison's method was to ask general knowledge questions.

> Of course [he continued], I don't care directly whether a man knows the capital of Nevada, or the source of mahogany, or the location of Timbuktu. Of course I don't care whether he knows who Desmoulins and Pascal and Kit Carson were. But if he ever knew any of these things and doesn't know them now, I *do* very much care about that in connection with giving him a job. For the assumption is that if he has forgotten these things he will forget something else that has direct bearing on his job.

The questionnaire led Edison to two conclusions. The first was that education was failing badly in its job. Of the initial 718 men tested, only 57 could reach even the standard which he rated as "fair." The second conclusion was that the questionnaire worked. He normally employed only those successful in the test, and all of them made good executives. Occasionally he would take others; but they made poor executives and as far as Edison was concerned that settled it.

A good account of the system at work has been given by A. L. Shands.

> We were taken to the third floor of the laboratory, seated at a long, unfinished wooden table, and presented with the questions, mimeographed and with intervening spaces for our answers. We had till noon. That gave us a bit less than two and a half hours for the 150 questions.

Those which demanded technical data or chemical formulae, could be answered with no more than a routine knowledge of current textbooks. Some were mathematical puzzles. Others asked who wrote this or who discovered that. A few demanded judgment such as: "If you were a salesman and you met a prospective customer out with a chorus girl, would you tell his wife?" One dealt with poker, and it appeared to Shands that an acceptable answer was: "I don't play poker."

Shands got a job as inspector, a freelance employee who was allotted a section of the factory in which he was expected to act as Edison's eyes and ears, make ideas for improving production, and report loafing or inefficiency. "The usual life of an inspector used to be about thirty days," says Shands. "By that time he was sick of making enemies, or realized a shortage of ideas. He quit."

But in the 1920s there was an unending stream of young men anxious to try their hand at being Edison inspectors. This was natural enough. The uncritical adulation with which he was now greeted everywhere, the justifiable importance given to his successes, and the ease with which his failures were, like his more extravagant claims, glossed over or forgotten, increased his guru status and to work within his orbit was something of a triumph. A forerunner of the contemporary T.V. sage, eager to rise to any question, Edison had become one of the most accessible targets for interviewers. Once anxious to elaborate on the future of his latest invention, he was now prepared to tackle the troubles of the world and of humanity, and usually just as ready with a solution.

He still saw into the future with considerable skill, foreseeing the problem of traffic in great cities, the shorter working day, and the leisure problem that it would create. In 1922 he was more percipient than most men—and most scientists—on the potential of atomic energy. Rutherford had chipped—but not split—the atom only three years before Edison wrote of atomic energy in his diary: "It may come some day. As a matter of fact, I am already experimenting along the lines of gathering information at my laboratory here.... So far as atomic energy is concerned there is nothing in sight just now. Although tomorrow some discovery might be made."

He still had an inquiring interest in "everything" and astonished one reporter by telling him that he had started a major investigation of the ether.

> I'm doing this for recreation [he said]. I am just reading everything on the subject I can find, and finding out what the other fellow has thought and published and discovered. I am not at all satisfied that the present-day thought on the ether is correct. If Einstein can find something new to say about space and time and geometry it may be the last word hasn't been said on the ether. But I couldn't say a word about it now. I haven't data enough. I can't work without data—*all* the data.

At the age of seventy-four; as a hobby; and the need for *all* the data before he could begin!

His optimism was an irresistible attraction. Asked to comment on America's economic situation by a *New York Times* reporter, he opened up with: "Are you one of the howlers, too?" There was no doubt that economic conditions in the United States were, as he put it, "somewhat upset". But they were, he went on, "not so seriously disarranged that we cannot remedy them by grit, determination and hard work.... Don't call it a panic. It is nothing but a period of depression, and nothing to worry over, provided we set ourselves resolutely to the task of overcoming it." On such subjects he would be given serious attention although his plan for "new money," to be paid to farmers and similar producers lodging their goods in government warehouses, was summarily dismissed by the experts as outside the realm of practical banking.

As with most sages who provide panaceas for everything, Edison's words of wisdom were sometimes shot through not only with common sense but with banality. At times there was also a suspicion of tailoring his words to his audience, particularly when he spoke about religion, a field where his inclination toward the truth fought a constant battle with his disinclination to offend.

For his first sixty years he appears to have given little time and less thought to the subject and only in 1910 did he startle many Americans with his statement that a personal God meant nothing to him. Almost twenty years later, asked what the word "God" meant to him, he replied "nothing," an admission erased by Mrs. Edison who hastily explained that her husband was referring solely to the word and not to the idea it represented. In 1910 colleagues pointed out that unorthodox views were bad for business and for the next few years he carefully steered clear of religion. However, even before this there were indications that the occasional doubt crossed his mind. When a lightning flash lit up the laboratory where he was talking with Samuel Insull, he had commented: "That's the Opposition, Sammy"; then, as the thunder rolled away: "There's an Engineer—Somewhere." However, the occasional remark should not be taken too seriously and Edison's attitude is better represented by his refusal to leave the laboratory to attend church with his wife: "Can't go. But you go, and pray like hell that this experiment is successful."

A subtle change, temporary, and too indefinite to be described as a move from atheism to agnosticism, but a change nevertheless, came after 1918. The tidal wave of interest in spiritualism which followed the end of a war in which millions had died aroused his interest in the possibilities of an afterlife; so, no doubt, did the fact that he was himself already past the biblical three score years and ten.

Typically enough, it was the ouija board, a contrivance claimed to give messages from those who had "passed over," which first attracted his attention. It was, he felt, far too unscientific an instrument to help in psychic research;

234

something better was needed and he began to devise it. The inevitable followed, and before long French writers had Edison operating a telegraph office to the hereafter where those wishing to communicate with the dead could be serviced promptly and efficiently. He told an interviewer from the *Scientific American*:

> I don't claim that our personalities pass on to another existence or sphere. I don't claim anything because I don't know anything about the subject; for that matter, no human being knows. But I do claim that it is possible to construct an apparatus which will be so delicate that if there are personalities in another existence or sphere who wish to get in touch with us in this existence or sphere, this apparatus will at least give them a better opportunity to express themselves than the tilting tables and raps and ouija boards and mediums and the other crude methods now purported to be the only means of communication.

What he had in mind was an apparatus which operated like a valve. In a modern powerhouse, a man with his one-eighth of a horsepower could turn a valve which started a 50,000-horsepower steam turbine. In his apparatus, the slightest effort which it intercepted would be magnified so many times that it would produce a usable record. One of his collaborators, he revealed to the *Scientific American*, had recently died while working on the apparatus, and "he ought to be the first man to use it if he is able to do so."

However, nothing more was heard of Edison's attempt to get in touch with the next world and the public was left to judge his views by the Delphic, and often contradictory, statements made in interviews and articles. The most revealing was in the *Forum*, and although he failed to answer the self-imposed question, "Has Man an Immortal Soul?," he gave a clear résumé of his beliefs toward the end of a long life.

After praising the Sermon on the Mount, he confessed: "I cannot see that creeds amount to anything, and personally I am amazed because apparently sound minds set such great store by them." He could not be impressed by the idea that merely spoken prayers were likely to be answered but was certain that "lived prayers" would be.

> Once convince boys and girls and men and women that if they are not straight and square and honest, if they are not reasonably unselfish and inclined to follow the great precept of the Golden Rule, they cannot possibly be happy, and you will accomplish about all that really is necessary in the way of religious teaching.

As a man who had spent his life investigating the natural world he continued to wonder at "an oak leaf," "the busy efforts of a squirrel to lay up food for the winter," and "the infinite beauty of a snowflake," considering them more

235

inducive to revelation than "all the textbooks of the theological seminaries." When it came to institutionalized religions, he showed his hand.

> Some of the existing so-called religious creeds remind me of certain other savage theories which have been embalmed in law where they are generally accepted because they have been so accepted. We establish many rules and consider them inspired, when, in the light of actual knowledge, they are no more inspired than the rule of the wild headhunter who cannot get a really nice girl to marry him unless he gives to her two human heads.

On the ultimate question he was both clearer and more outspoken in private correspondence. To Joseph Lewis, after reading his *The Tyranny of God*, he wrote:

> I think as you do that death ends all, yet I do not feel certain, because there are many facts that seem to show that the real units of life are not the animal mechanism itself but groups of millions of small entities living in the visible cells—the animal being their mechanism for navigating their environment. And when the mechanism fails to function, i.e. dies, the groups go out into space to go through another cycle. The entities are each highly organized and perform their allotted tasks. If there is anything like this we still have a fighting chance.

As reluctant as ever to turn away an interviewer, still his own hard taskmaster ordering himself a long day at the West Orange works, Edison remained almost totally unchanged by the tributes and honors of the 1920s. The Congressional Gold Medal was satisfying, but it was set against the blaze of electric light as one approached New York or any other big city. Election to the American Academy of Sciences was a fitting tribute, even though it came late in the day; yet it was half a century earlier that the prototype of the talking machine, now sold by the tens of thousands, had "presented itself" to the Academy.

Like most men past an active seventy-five, Edison had become a trifle set in his ways, following an almost unbreakable routine that began with an eight o'clock breakfast of half a grapefruit, coffee, and a slice of toast. He tended to be pernickety about small things and if any member of the family had previously taken a glance at the *New York Times*, carefully folded beside his plate, then there was trouble. "It has to be in perfect order," Mrs. Edison confessed, "just as delivered by the newsdealer. If the pages have been disarranged or if there is any indication that anybody has been reading it before he sees it, Mr. Edison wants to know who has been interfering with his *Times*. It is the nearest he ever comes to a storm."

The quick breakfast was followed by a short drive down to the factories, study of the latest reports from men working on any special project, then on to what-

ever investigations he himself had in hand. It was usually late in the day before
he got back home although occasionally he took an afternoon drive in the car.
There were business trips to New York or Chicago, but what he still avoided,
with all the vigor of his earlier days, was the set-piece formal dinner. His deafness
prevented him from hearing the speeches, and he still hated the business of dressing
up.

Like most aging men, he regretted some things. In particular there were
the first commercial talkies.

> They have spoiled everything for me [he admitted]. There isn't any more good
> acting on the screen. My, my, how I would like to see Mary Pickford or Clara
> Bow in one of those good oldfashioned silent pictures. They concentrate on the
> voice now; they've forgotten how to act. I can sense it more than you because
> I am deaf. It's astounding how much more a deaf person can see.

His tastes were as unsophisticated as when he had been an innocent tramp-
telegraphist and if a picture ended without the prospect of hero and heroine
living happily ever after, he was severely critical.

He continued to take a no-nonsense attitude to his habits and his health.
Only with difficulty was he persuaded, at the age of eighty, to visit a doctor.
When an even more rigid diet was proposed he then treated himself as the subject
in one of his own experiments, carefully analyzing the effects of different foods
upon his sytem. Having cut out every item that seemed to cause trouble, he
announced that he was going on a milk diet. When, at the age of eighty-one,
he developed pneumonia he insisted on continuing work, refused to take medi-
cine and stoutly maintained that he could sleep everything away.

The work, remarkably enough, was for a completely new project on which
he had embarked at the age of eighty. A decade earlier, in 1915, touring Burbank's nurseries at Santa Barbara, he had discussed with Ford the problem of
rubber supplies if America was drawn into the European war. Seven years later,
during the summer camping holiday, both Ford and Firestone, aware that Britain's huge Malayan rubber forests gave her almost a monopoly of supplies,
suggested that America's greatest inventor should turn his mind to the situation.
Another five years passed and it was only in 1927 that the Edison Botanic Research Company was founded. Ford and Firestone each put up more than
$90,000 with which Edison set up a new laboratory at Fort Myers, complete
with fields where potential sources of supply could be grown. His enthusiasm
now was generated by fear of another war.

> Don't make any mistake about that war [he wrote]; it will come. We may run
> along for a good many years without it, but sooner or later the nations of Europe

will combine against the United States. The first thing they will do will be to cut off our rubber supply.

From Fort Myers botanists were dispatched—much as they had been sent to find bamboo for electric light filaments—to countries throughout the world. Their task was to find plants from which rubber latex could be extracted. The plants had to be quick-growing since, as Edison pointed out, supplies would be needed within a year of war breaking out, while in peacetime the country could not be expected to use valuable agricultural space for rubber year after year. The ideal would thus be

> an annual crop, something which the farmer can sow in the field by machinery, which will come to maturity in eight or nine months, which can then be harvested by machinery, and from which rubber can be obtained by processes almost entirely mechanical, with the least amount of hand labor.

It was a difficult enough aim and in Mrs. Edison's words: "Everything has turned to rubber in our family. We talk rubber, think rubber, dream rubber. Mr. Edison refuses to let us do anything else." He did not lock his family indoors until they had helped find the solution, but it is not impossible that the idea crossed his mind.

Within a year more than 3,000 plants had been collected and about 200 were found to contain rubber latex. The following year, after examination of 14,000 plants, Edison decided that a variety of goldenrod offered the solution. Cross-breeding had produced a plant more than twelve feet high and capable of giving respectable quantities of latex, and from a shipment which he sent to Firestone four tires were made for a Ford tourer. But one intransigent fact remained; rubber made from goldenrod was wildly uneconomic. Nevertheless, Edison, no stranger to the problems of economics, was confident that production costs could be whittled down so that goldenrod could compete with imported rubber. This might have been the case had he been ten years younger. But now, in his early eighties, he fell ill with a combination of digestive and kidney troubles. For the time being at least, the problem of America's indigenous rubber supplies had to be laid aside. When it was finally taken up once more by Edison's successors it was synthetic rubber, not goldenrod, which offered the quickest route to self-sufficiency.

By the autumn of 1929 he was well enough to take part in an historic ceremony engineered by Henry Ford. A decade previously Ford, giving evidence in a libel suit against the *Chicago Tribune*, had delivered the judgment that "history is bunk." He was not allowed to forget the phrase and during the following years planned to counterbalance the error.

When Ford had finished with it, history would no longer be a subject entombed in dull books; instead, it would be learned by inspecting the objects that were part of history. To his headquarters at Dearborn he therefore began to gather the machinery and equipment that was part and parcel of America's story. To Dearborn he brought a replica of the Wright brothers' bicycle shop at Dayton, Ohio, where they had fashioned the craft that made men airborne. He brought Luther Burbank's office from Santa Rosa. Then it struck him that 21 October 1929 would mark the fiftieth anniversary of the incandescent lamp. General Electric was already planning celebrations at their Schenectady headquarters; but Edison had barely been consulted by them and he was happy to collaborate with his old friend when Ford proposed that a "Golden Jubilee of Light" should be held at Dearborn. As part of the celebrations, there would be opened his Museum of History in which a star exhibit would be a reproduction of Edison's laboratory at Menlo Park.

In September 1928 Edison watched the foundation stone of the museum being laid, formally left his footprint in soft concrete, and signed it with his name and date. Thirteen months later a few days before 21 October, he returned, to be astounded as he entered the quadrangle in which the reconstructed laboratory stood, closed in by a white picket fence indistinguishable from the fence hammered in round the Menlo Park laboratory more than half a century earlier. The exhibit was a masterpiece of its kind, and as Edison looked at the soil, carloads of which had been brought from Menlo Park itself, he noted: "H'm, the same damn old New Jersey clay."

Inside, he led the way into the re-created second-floor laboratory, accompanied by Ford, and one of the few remaining survivors from the Menlo Park days: Francis Jehl, found in Europe and shipped back across the Atlantic to play a supporting role in the operation.

> As [Edison] walked to a chair and sat down, his companions in the party remained where they stood, apart from him a dozen feet [wrote a reporter from the *Detroit Free Press*]. No word was spoken; it was as if by common consent the spectators instinctively felt awed here, in the presence of an old man upon whom the memories of eighty-two years were flooding back. He sat there, silent, his arms folded, an indescribably lonely figure, lonely in the loneliness of genius, of one who, somehow, has passed the others, who no longer has equals to share his world, his thoughts, his feelings.
>
> For five, perhaps ten, minutes, the scene was unmarred by a word or a gesture, except that now and then Edison looked about him and his eyes dimmed. Suddenly he cleared his throat and the spell was broken.

Meticulous attention to detail was evident and when Ford handed Edison the old mortar he had used, reconstructed from scrap found on the rubbish heap,

Edison commented that the whole building, and its contents, was nine-tenths perfect. Ford, nettled, asked what was wrong. "Our floor," replied Edison, "was never as clean as that."

On the morning of the 21st, when a special anniversary stamp appeared showing the original sewing-thread bulb, President and Mrs. Hoover arrived in Dearborn. They were met by Edison and his wife and the assembled company then traveled on a nineteenth-century train, drawn by a wood-burning locomotive. During the brief journey he took a train-boy's basket and calling "candy, newspapers" in a weak voice offered his wares to the company.

After the President had helped an obviously aging Edison from the train the distinguished visitors toured Ford's Museum of History. That evening they returned to the laboratory for the set-piece event. With the lights lowered, the main guests crowding into the laboratory, and millions throughout America listening to the broadcast commentary, Edison prepared to reenact a modified version of the drama which had taken place half a century ago.

Francis Jehl prepared the vacuum pump. Edison, sitting in a chair beside it, eventually ordered: "Go ahead, Francis." Then, as the pressure fell, he rose and connected the wires. The lamp began to glow, glowed brighter, then burst into light. At that moment lamps throughout the area blazed up, and in scores of cities across America those who had dimmed their lights in tribute to genius turned them on to blazing full strength.

But the genius still had to endure the ensuing banquet. At the door of the hall he almost collapsed and was induced to go on only by Mrs. Edison. The seats of honor at the top table had been reserved for President and Mrs. Hoover. They insisted that the Edisons sit there instead. Around them were some 500 guests, a carefully selected sample from *Who's Who*, while that evening there were congratulatory messages from the Prince of Wales in England, from President von Hindenburg in Germany, and from Commander Richard Byrd, sitting out the Antarctic climate of Little America. And from Germany, also, there was relayed the voice of Albert Einstein.

Edison spoke briefly, ending with a tribute to Henry Ford: "I can only say to you that in the fullest and richest meaning of the term—he is my friend. Good night." As expected, he had stayed the course. But now he collapsed, recovering only after resting in a nearby room and an injection of adrenalin.

After Dearborn Edison began to fail, although there is no evidence that the strain, physical and emotional, had any part in this. He visited the factories less frequently and for shorter periods. When he worked it was not in his old laboratory but in a room at Glenmont. The afternoon car rides became shorter, his grip on affairs less sure. Throughout 1930 and into 1931 he clung on, only reluctantly allowing the reins to be taken from his hands. The old spirit was still there and in June 1931, with America in the deepest trough of the depression, he sent to a lighting convention in Atlantic City a message breathing the old optimism.

> My message to you is to be courageous. I have lived a long time. I have seen history repeat itself again and again. I have seen many "depressions" in business. Always America has come out strong and more prosperous. Be as brave as your fathers were before you. Have faith—go forward.

It was his last public message. On 1 August he collapsed and his doctor diagnosed a medley of troubles that included Bright's disease, uremic poisoning, and diabetes. He was expected to last only a few days and a statement from his doctor that "Mr. Edison might be compared to a ship sailing down a narrow channel. He may steer a safe course or he may strike a rock," did little to reassure an anxious nation. But after a critical week no more daily bulletins were necessary, and a little later the afternoon automobile rides were resumed. Once more it appeared that the experts had been wrong.

Early in September there was a relapse. Once again Edison fought back, questioning his doctors about the chemical effects of the medicines they were giving him, discouraged only by the fact that he was not recovering as quickly as he had expected. Early in October the Pope sent the first of two messages asking for news of his health and the Chamber of Commerce in Fort Myers agreed to hold Sunday, 4 October, as a day of prayer for his recovery.

Nine days later he passed into a coma, but not before he had looked from his bedroom window down across the valley where he had played with his children in the early years of the century, and smilingly remarked: "It is very beautiful over there"; and not before he had received a last visit from Firestone to whom he indicated, with a look of triumph, four vulcanized specimens of goldenrod rubber.

He died on 18 October. His doctor, recalling his description of "over there," asked if the great inventor had "climbed the heights which lead into eternity and caught a glimpse of the veil which obstructs our earthly vision?" In the ensuing argument on Edison's beliefs the president of the Freethinkers of America produced photostats of his annual dues to the society. An honest family statement, trying to deal with the claims that he had changed his views, noted that "this is a difficult question to answer because of the widespread misunderstanding of what his beliefs actually were."

As Edison's body lay in state in the laboratory at West Orange, suggestions for a dramatic gesture marking his achievements poured into his home and into the White House. Emil Ludwig, the German biographer who had more than once praised his genius, proposed that lights all over the world should be ex-tinguished for a symbolic minute. Less ambitiously, it was proposed that Presi-dent Hoover should order all electric current to be turned off throughout the United States for a minute on the day of the funeral. When the impracticability of such ideas was appreciated a voluntary darkening of all but essential lights was suggested. The result was impressive.

At 6:59 P.M. Pacific time on 21 October, as the sun went down off California, and in Denver, where it was 7:59 mountain time, the lights went out. It was 8:59 in Chicago and here the elevated trains and the streetcars stopped for a minute, the lights were extinguished in the city, and all the Mississippi Valley from Cairo to the sea was in darkness. It was 9:59 eastern standard time in New York and all but essential traffic signs were switched off, the lights on Broadway were dimmed and the torch on the Statue of Liberty was extinguished.

For a minute America was almost back in the age of kerosene and gas lamps. Then, from coast to coast, the electric light blazed out again.

"Lights Out," from the New York World-Telegram *after Edison's death*

Bibliography

ACHESON, Edward Goodrich. *Edward Goodrich Acheson: A Pathfinder, Inventor, Scientist, Industrialist.* Port Huron, Mich.: Acheson Industries, 1965

ACKERMAN, Carl W. *George Eastman.* Clifton, N.J.: Kelley, 1930; London: Constable, 1930

ALLISTER, Ray. *Friese-Greene: Close-up of an Inventor.* New York: Arno Press, 1948; London: Marsland Publications, 1948

BAKER, E. C. *Sir William Preece, F.R.S., Victorian Engineer Extraordinary.* London: Hutchinson, 1976

BENSON, A. L. "Wonderful New World Ahead of Us." *Cosmopolitan Magazine,* February 1911

BERNHARDT, Sarah. *My Double Life: Memoirs of Sarah Bernhardt.* New York: Appleton-Century, 1907; London: Heinemann, 1907, Peter Owen, 1977

BISHOP, William H. "A Night With Edison." *Scribner's Monthly,* November 1878

BRYAN, George S. *Edison: The Man and His Work.* New York and London: Knopf, 1926

BURBANK, Luther. *The Harvest of the Years.* New York: Houghton-Mifflin, 1927; London: Constable, 1927

BURLINGAME, Roger. *Engines of Democracy: Inventions and Society in Mature America.* New York: Scribner's, 1940

CASSON, Herbert N. *The History of the Telephone.* Chicago: A. C. McClurg, 1910

CLARK, Victor S. *History of Manufactures in the United States, 1860–1914.* Washington, D.C.: Carnegie Institute of Washington, 1928

DE LONG, Emma, ed. *The Voyages of the Jeannette: The Ship and Ice Journals of George W. De Long.* New York: Houghton Mifflin, 1883; London: Kegan Paul, Trench, 1883

DICKSON, W. K. L., and DICKSON, Antonia. *The Life and Inventions of Thomas Alva Edison.* New York: Crowell, 1894; London: Chatto & Windus, 1894

DICKSON, W. K. L., and DICKSON, Antonia. *History of the Kinetograph, Kinetoscope and Kinetophonograph.* New York: Arno Press and the New York Times, 1895, 1970

DUNLAP, Jr., Orrin E. *Marconi: The Man and His Wireless.* New York: Macmillan, 1937

DYER, Frank Lewis, and MARTIN, Thomas Commerford. *Edison: His Life and Inventions.* New York and London: Harper & Brothers, 1910

FLEMING, Sir John Ambrose. *Memories of a Scientific Life.* London: Marshall, Morgan & Scott, 1934

FLEMING, J. A. *Fifty Years of Electricity: The Memories of an Electrical Engineer.* New York and London: The Wireless Press, 1921

FORD, Henry, with CROWTHER, Samuel. *Edison as I Know Him.* New York: Cosmopolitan Book Corporation, 1930; London (as *My Friend, Mr. Edison*): Ernest Benn, 1930

GARBIT, F. J. *The Telephone—Edison's Speaking Phonograph.* In Dana Estes, ed. *Half-Hour Recreations in Popular Science,* Series Two. Boston: 1884

GELATT, Roland. *The Fabulous Phonograph: The Story of the Gramophone from Tin Foil to High Fidelity.* New York: Appleton-Century, 1966; London: Cassell, 1956

GRODINSKY, Julius. *Jay Gould, His Business Career, 1867–1892.* Philadelphia, Pa.: University of Pensylvania Press, 1957

HAMMOND, John Winthrop. *Men and Volts: The Story of General Electric.* New York: Lippincott, 1941

HARLOW, Alvin F. *Old Wires and New Waves: The History of the Telegraph, Telephone and Wireless.* New York: Appleton-Century, 1936

HENDRICKS, Gordon. *The Edison Motion Picture Myth.* Berkeley, Cal.: University of California Press, 1961

HERNDON, Booton. *Ford: An Unconventional Biography of the two Henry Fords and Their Times.* New York: Weybright, 1969; London: Cassell, 1970

HOPKINSON, John. *Original Papers of the late John Hopkinson, D.Sc., F.R.S.* with memoir by B. Hopkinson. London: Cambridge University Press, 1901

HORSCHITZ, A., and OESTREICH, Paul. *Edison and His Competition: A Critical Study.* London: W. & G. Foyle, 1932.

ILES, George. *Inventors at Work.* New York: Doubleday, Page & Co., 1906

INSULL, Samuel. *Central Station Electric Service.* Chicago: privately printed, 1915

INSULL, Samuel. *Public Utilities in Modern Life.* Chicago: privately printed, 1924

JEHL, Francis. *Reminiscences of Menlo Park.* Dearborn, Mich.: Edison Institute, 1937–9

JEWKES, John; SAWYERS, David; and STILLERMAN, Richard. *The Sources of Invention.* New York and London: Macmillan, 1958

JOHNSON, Robert Underwood. *Remembered Yesterdays.* Boston: Little, Brown, 1923

JONES, Francis Arthur. *Thomas Alva Edison: An Intimate Record.* New York: Crowell, 1907; London: Hodder & Stoughton, 1907

JOSEPHSON, Matthew. *Edison.* New York: McGraw-Hill, 1959; London: Eyre and Spottiswoode, 1961

JOSEPHSON, Matthew. *The Robber Barons: The Great American Capitalists, 1861–1911.* New York: Harcourt Brace Jovanovitch, 1962; London: Eyre and Spottiswoode, 1962

LATHROP, George Parsons. "Talks with Edison." *Harper's Monthly Magazine*, February 1890

LIEF, Alfred. *Harvey Firestone: Free Man of Enterprise*. New York: McGraw-Hill, 1951

LYND, William. *Edison and the Perfected Phonograph*. Tunbridge Wells: A. K. Baldwin, n.d.

MACDONNELL, Kevin. *Eadweard Muybridge: The Man Who Invented the Moving Picture*. Boston: Little, Brown, 1972; London: Weidenfeld & Nicolson, 1973

MCCLURE, J. B., ed. *Edison and His Inventions, Including the Many Incidents, Anecdotes and Interesting Particulars Connected with the Life of the Great Inventor*. Chicago: Rhoades and McClure, 1879

MCDONALD, Forrest. *Insull*. Chicago and London: University of Chicago Press, 1962.

MARCOSSON, I. F. "The Coming of The Talking Picture." *Munsey's Magazine*, March 1913

MARTIN, T. Commerford. *Forty Years of Edison Service: 1882–1922*. New York: The New York Edison Co., 1922

MEADOWS, A. J. *Science and Controversy: A Biography of Sir Norman Lockyer*. Cambridge, Mass.: M.I.T. Press, 1972; London: Macmillan, 1972

MILLER, F. T. *Thomas Alva Edison: Benefactor of Mankind; The Romantic Life Story of the World's Greatest Inventor*. Philadelphia, Pa.: John C. Winston, 1931; London: Stanley Paul, 1932

MILLER, Raymond C. *Kilowatts at Work: A History of the Detroit Edison Company*. Detroit, Mich.: Wayne State University Press, 1957

MUYBRIDGE, Eadweard. *Animals in Motion*. London: Chapman & Hall, 1899

NERNEY, Mary C. *Thomas A. Edison: A Modern Olympian*. New York: Harrison Smith & Robert Haas, 1934

NEVINS, Allan. *Ford: The Times, the Man, the Company*. New York and London: Charles Scribner's Sons, 1954

NEVINS, Allan and HILL, Frank Ernest. *Ford: Expansion and Challenge, 1915–1933*. New York and London: Charles Scribner's Sons, 1957

O'DEA, William T. *The Social History of Lighting*. London: Routledge and Kegan Paul, 1958

O'NEILL, John. *Prodigal Genius: The Life of Nikolas Tesla*. New York: Ives, Washburn, Inc., 1944; London: Spearman, 1968

PAINE, Thomas. *The Works of Thomas Paine*. New York: W. H. Wise, 1934

PASSER, Harold C. *The Electrical Manufacturers, 1875–1900*. Cambridge, Mass.: Harvard University Press, 1953

PREECE, W. H. *The Phonograph: or Speaking Machine*. London: London Stereoscopic Co., n.d.

RAMSAY, M. L. *Pyramids of Power: The Story of Roosevelt, Insull and the Utility Wars*. Indianapolis and New York: Bobbs-Merrill, 1937; reprinted New York: Da Capo, 1975

RAMSAYE, Terry. *A Million and One Nights: A History of the Motion Picture*. New York: Simon & Schuster, 1964; London: Frank Cass, 1964

READ, Oliver, and WELCH, Walter L. *From Tin Foil to Stereo: Evolution of the Phonograph*. Indianapolis and New York: Sams and Bobbs-Merrill, 1959

RHEINHARDT, E. A. *The Life of Eleanor Duse*. New York: Blom, 1930; London: Secker, 1930

ROBERTSON, J. H. *The Story of the Telephone*. London: Science Book Club, 1948

ROSANOFF, M. A. "Edison in His Laboratory." *Harper's Magazine*, September 1932

RUNES, Dagobert D., ed. *The Diary and Sundry Observations of Thomas Alva Edison*. New York: Philosophical Library, 1948

SATTERLEE, Herbert L. *J. Pierpont Morgan: An Intimate Portrait*. New York: Macmillan, 1939; reprinted New York: Arno Press, 1975

SCOTT, J. D. *Siemens Brothers, 1858–1958*. London: Weidenfeld & Nicolson, 1958

SCOTT, Lloyd N. *Naval Consulting Board of the United States*. Washington, D.C.: Government Printing Office, 1920

SHANDS, A. L. *The Real Thomas Edison*. Girard, Kans.: Haldeman-Julius Publications, 1929

SHAW, George Bernard. *The Irrational Knot*. London: Constable, 1931

SHAW, G. M. "Sketch of Thomas Alva Edison." *Popular Science Monthly*, August 1878

SIMONDS, William Adams. *Edison: His Life, His Work, His Genius*. London: Allen & Unwin, 1935

SMITH, Albert. *Two Reels and a Crank*, New York: Doubleday, 1952

SORENSEN, Charles E., with WILLIAMSON, Samuel T. *My Forty Years with Ford*. London: Cape, 1957; New York: Collier Macmillan, 1962

SPRAGUE, Harriet. *Frank J. Sprague and the Edison Myth*. New York: William-Frederick Press, 1947

SWAN, Mary E., and SWAN, Kenneth R. *Sir Joseph Wilson Swan, F.R.S.* London: Ernest Benn, 1929

TATE, Alfred O. *Edison's Open Door*. New York: Dutton, 1938

THOMPSON, Holland. *The Age of Invention: A Chronicle of Mechanical Conquest*. Yale Chronicles of America, Vol. 39. New York: U.S. Publishers' Association, n.d.; London: Oxford University Press, 1921

THOMPSON, Jane Smeal, and THOMPSON, Helen G. *Silvanus Phillips Thompson, D.Sc., LL.D., F.R.S.: His Life and Letters*. London: Fisher Unwin, 1920

UPTON, Francis R. "Edison's Electric Light." *Scribner's Monthly*, February 1880

VANDERBILT, Byron M. *Thomas Edison, Chemist*. Washington, D.C.: American Chemical Society, 1971

WARREN, Waldo P. "Edison on Invention and Inventors." *Century Illustrated Monthly Magazine*, July 1911

WHITE, Trumbull. *The Wizard of Wall Street and his Wealth, or the Life and Deeds of Jay Gould*. Chicago: Mid-Continent Publishing Co., 1892

WOODBURY, David O. *Beloved Scientist—Elihu Thomson, A Guiding Spirit of the Electrical Age*. New York: McGraw-Hill, 1944

YOUNG, J. Lewis. *Edison and the Phonograph*, New York: Edison Phonograph Co., 1890

Among the journals and newspapers consulted are *Science, Scientific American, Nature*, the *Operator and Electrical World*, the *Electrical Engineer, Harper's, Scribner's*, the *New York Times, New York Herald, New York Tribune* and *New York World*, the London *Times* and the Bulletins of the Edison Electric Light Company.

References

Full details of the sources quoted are given in the Bibliography, pages 244–45.

Page
9 I found out: *New York World*, quoted Francis Arthur Jones, *Edison: An Intimate Record* (London), p. 8 (afterwards referred to as "Jones")
9 It was a revelation: Edison, Introduction to *The Works of Thomas Paine*
10 I at once came to the conclusion: W. K. L. Dickson and Antonia Dickson, *The Life and Inventions of Thomas Alva Edison* (London), p. 6 (afterwards referrred to as "Dickson")
10 I am not a mathematician: Dagobert D. Runes, ed., *The Diary and Sundry Observations of Thomas Alva Edison*, p. 45 (afterwards referred to as "Runes")
10 I can always hire: Frank Lewis Dyer and Thomas Commerford Martin, *Edison: His Life and Inventions* (London), p. 26 (afterwards referred to as "Dyer and Martin")
11 he would probably have become: A. Horschitz and Paul Oestreich, *Edison and His Competition: A Critical Study*, p. 47
12 Ma! I'm a bushel of wheat: George Parsons Lathrop, "Talks With Edison," *Harper's Monthly Magazine*, February 1890 (afterwards referred to as "Lathrop")
12 mind was an electric thunderstorm: J. Lewis Young, *Edison and the Phonograph*, p. 6 (afterwards referred to as "Young")
12 After being on the train: quoted Holland Thompson, *The Age of Invention: A Chronicle of Mechanical Conquest* (London), p. 205
12 I ran after it: Henry Ford with Samuel Crowther, *My Friend, Mr. Edison* (London), p. 22 (afterwards referred to as "Ford with Crowther")
13 While I could hear unerringly: Runes, p. 48
13 Deafness, pure and simple: Runes, p. 53
14 My copy: Lathrop
14 I grasped the situation: Lathrop
15 I took my fifteen hundred papers: Lathrop
16 The train: Dickson, p. 15
18 A classic example: Byron M. Vanderbilt, *Thomas Edison, Chemist*, p. 18 (afterwards referred to as "Vanderbilt")
19 A memorable experience: Robert Underwood Johnson, *Remembered Yesterdays*, p. 59 (afterwards referred to as "Johnson")
19 He was one of: Lathrop
20 I got two old: *Scientific American*, 1 April 1905

21 Down in Virginia: A. L. Benson, "Wonderful New World Ahead of Us," *Cosmopolitan Magazine*, February 1911
23 Soon the New York operator: Dickson, p. 41
24 Adams: George S. Bryan, *Edison: The Man and His Work*, p. 49 (afterwards referred to as "Bryan")
24 To stop is to rust: Francis Jehl, *Reminiscences of Menlo Park*, p. 465 (afterwards referred to as "Jehl")
24 A harvest: Jehl, p. 465
25 Young man: Lathrop
29 To draw the long bow: Matthew Josephson, *Edison* (London), p. 83
30 How would $40,000: Dyer and Martin, p. 132
30 a bloated: Edison to parents, 30 October 1870
31 I kept only payroll accounts: *Scientific American*, 27 December 1902
31 Mr. Edison had his desk: *New York Times*, 19 October 1931
49 I came in one night: J. B. McClure, ed., *Edison and His Inventions*, p. 17 (afterwards referred to as "McClure")
50 I got my apparatus down: Dyer and Martin, p. 151
50 The English: Johnson, p. 115
51 This problem: Dyer and Martin, p. 156
51 the story of the fight: Alvin F. Harlow, *Old Wires and New Waves: The History of the Telegraph, Telephone and Wireless*, p. 405 (afterwards referred to as "Harlow")
52 I had picked: Dyer and Martin, p. 158
53 His conscience: Dyer and Martin, p. 164
53 because he was so able: Dyer and Martin, p. 164
53 In it: Harlow, p. 406
54 Orton told me: E. C. Baker, *Sir William Preece, F.R.S., Victorian Engineer Extraordinary*, p. 157 (afterwards referred to as "Baker")
54 Everybody steals: M. A. Rosanoff, "Edison in His Laboratory," *Harper's Magazine*, September 1932 (afterwards referred to as "Rosanoff")
54 I spent the whole: *Scientific American*, 2 April 1892
54 Inasmuch as every mile: *Scientific American*, 2 April 1892
56 With a delicately constructed machine: *Scientific American*, 5 September 1874
57 I constructed: McClure, p. 101
58 My ambition: Dyer and Martin, p. 180
58 With all the bulk: Herbert N. Casson, *The History of the Telephone*, p. 59 (afterwards referred to as "Casson")
59 How to compete: Casson, p. 69

60 of such stentorian: George Bernard Shaw, Preface vi, *The Irrational Knot*
60 Finally: *New York Tribune*, 1 September 1879
61 His quaint: *New York Tribune*, 1 September 1879
61 It is a common complaint: *The Times*, 10 May 1879
61 A gentleman: *The Times*, 10 November 1879
62 Their gangs: Baker, p. 180
62 These deluded: George Bernard Shaw, Preface vii, *The Irrational Knot*
63 not altogether in accordance: J. H. Robertson, *The Story of the Telephone*, p. 19
64 One letter: Runes, p. 32
65 I might invent: *Scientific American*, 2 April 1892
66 When I first noticed it: quoted *New York Times*, 19 October 1931
67 I had rented: Dyer and Martin, p. 267
67 When the public: Dickson, p. 101
67 I do not regard myself: *Scientific American*, 8 July 1893
69 when evening came on: Jehl, quoted Ford with Crowther, p. 82
69 He could go to sleep: Alfred O. Tate, *Edison's Open Door*, p. 143 (afterwards referred to as "Tate")
69 It often happened: Jehl, quoted Ford with Crowther, p. 83
70 If the normal process: Jehl, p. 522
70 Another blazing day: Baker, p. 162
70 I remember Sir William Thompson: *Washington Post*, 19 April 1878
70 the characteristic features: quoted *The Electrician*, 1 June 1878
70 Some of the office employees: Jehl, quoted Ford with Crowther, p. 85
71 I never think: Waldo P. Warren, "Edison on Invention and Inventors," *Century Illustrated Monthly Magazine*, July 1911 (afterwards referred to as "Warren")
71 I cheerily assured him: Runes, p. 43
71 He knows a lot: Jehl, p. 862
72 At one of the tables: *Scientific American*, 11 February 1911
72 I never attempted: Ford with Crowther, p. 86
73 By the simple inhabitants: Dickson, p. 104
74 was filled with theories: Lathrop
75 Will letter writing: *Scientific American*, 17 November 1877
75 to make the telephone: *Nature*, 29 November 1877
76 I didn't have much faith: Dyer and Martin, p. 208
77 I was never so taken aback: Dyer and Martin, p. 208
77 Mr. Thomas A. Edison: *Scientific American*, 22 December 1877
77 run by steam power: G. M. Shaw, "Sketch of Thomas Alva Edison," *Popular Science Monthly*, August 1878
78 I've made a good: *New York Graphic* quoted W. H. Preece, *The Phonograph or Speaking Machine*, p. 19
78 Two other things: *New York Graphic* quoted Preece, p. 20
78 These sheets: *New York World* quoted Preece, p. 22
79 You can take: *New York World* quoted Preece, p. 31
79 Mr. Edison showed: Roland Gelatt, *The Fabulous Phonograph: The Story of the Gramophone from Tin Foil to High Fidelity* (London), p. 28 (afterwards referred to as "Gelatt")
80 The main utility of the phonograph: Edison, "The Phonograph and its Future," The *North American Review*, June 1878
81 This tongueless: *Washington Post*, 19 April 1877
83 By holding a lighted cigar: Dickson, p. 112
83 The general excitement: *Electrical World and Engineer*, 5 March 1904
84 I freely showed: *New York Tribune*, 8 June 1878
85 My roommate was Fox: Dyer and Martin, p. 227
86 I hope you will come: A. J. Meadows, *Science and Controversy: A Biography of Sir Norman Lockyer* (London), p. 37
86 The ore: McClure, p. 124
87 Why can't the power: William Adams Simonds, *Edison: His Life, His Work, His Genius*, p. 135
87 Where waterpower: Runes, p. 52
88 so many others: Vanderbilt, p. 34
88 I believe I can beat you: quoted *Science*, 7 February 1947
88 His forehead: New York paper, quoted Jehl, p. 8
88 The accumulated wealth: *New York Herald*, 25 April 1879
89 I bought all the transactions: T. Commerford Martin, *Forty Years of Edison Service: 1882–1922*, p. 9 (afterwards referred to as "Martin")
89 Well, about 99 per cent: Samuel Insull, *Public Utilities in Modern Life*, p. 192 (afterwards referred to as "Insull")
90 We have an almost: *Scientific American*, 1 April 1905
91 Around October: *Electrical World and Engineer*, 5 March 1904
92 It is much after midnight: William H. Bishop, "A Night with Edison," *Scribner's Monthly*, November 1878
92 It is however: *The Telegraph Journal and Electrical Review*, 15 February 1879
93 I have just solved: Mary E. Swan and Kenneth R. Swan, *Sir Joseph Wilson Swan, F.R.S.*, p. 67
93 All that will be necessary: *New York Sun*, quoted *Nature*, 14 November 1878
94 He hasn't got: Rosanoff
94 noticed that he had: Insull, p. 124
95 Today I can see: quoted Dyer and Martin, p. 289
96 By turning (it): *New York Herald*, 27 April 1879
97 All night Batchelor: Jones, p. 154
97 I think we've got it: J. W. Hammond, *Men and Volts: The Story of General Electric*, p. 22
97 Edison was not seeking: Jehl, p. 522
99 Up to the present winter: *The Times*, 12 January 1880
99 We ate our supper: *The Times*, 14 January 1880
99 I shall never forget: *The Times*, 14 January 1880
99 For more than a week: *New York Herald*, 4 January 1880
100 After the electric light: *New York Herald*, 4 January 1880
103 There you have it: Jehl, p. 467
104 I don't know when: *New York World*, quoted *Nature*, 5 December 1878
105 I was then investigating: *Scientific American* Supplement No. 1480
105 No hero of mythology: *New York Evening Sun*, 2 May 1889
107 This will be: *Science*, 10 July 1880

107 that it is a very poor investment: *Science*, Vol. 105, 7 February 1947
108 In the parlor: *New York Herald*, 18 November 1880
108 I do not worry: *New York Herald*, 18 November 1880
109 A huge map: Philip A. Lange, speaking Engineers Club, Manchester, 15 March 1907, quoted Harriet Sprague, *Frank J. Sprague and the Edison Myth*, p. 12
110 The only answer: David O. Woodbury, *Beloved Scientist–Elihu Thomson, A Guiding Spirit of the Electrical Age*, p. 153 (afterwards referred to as "Woodbury")
110 Now that I have a machine: *The Electrician*, 2 November 1878
111 a central electricity lighting plant: *Nature*, 2 July 1896
111 The results of this stew: Jehl, p. 723
112 a most ingenious contrivance: *Nature*, 12 February 1880
112 What appeared to be: *Scientific American*, 11 February 1911
129 There in the end of the room: *Scientific American*, 11 February 1911
129 I engaged a man and a boy: *Scientific American*, 11 February 1911
130 irretrievably worthless: Emma De Long, ed., *The Voyage of the Jeannette: The Ship and Ice Journals of George W. De Long*, p. 154 (afterwards referred to as "De Long")
130 The electric machine: De Long, p. 163
131 A cellar was dug: Herbert L. Satterlee, *J. Pierpont Morgan: An Intimate Portrait*, p. 207
132 I cannot tell: Sarah Bernhardt, *My Double Life: Memoirs of Sarah Bernhardt* (London), p. 375
135 Insull looked at Edison: Forrest McDonald, *Insull*, p. 20
135 Whatever you do, Sammy: quoted *Chicago Daily Tribune*, 2 November 1934
135 Right after dinner: Insull, p. 332
136 It occurred to me: Martin, p. 34
136 a decided success: *New York Herald*, 22 December 1880
137 By working night and day: Dyer and Martin, p. 325
137 a large cone-shaped pendant: J. A. Fleming, *Fifty Years of Electricity: The Memories of an Electrical Engineer*, p. 159
138 Many unkind things: *Journal of the Society of Arts*, 16 December 1881
138 It is necessary to make: John Hopkinson, *Original Papers of the late John Hopkinson, D.Sc., F.R.S.*, p. xlii
139 A hundred dollars: Vanderbilt, p. 71
139 It was not until: *New York Times*, 5 September 1882
140 I replied: *New York Times*, 12 September 1922
140 Then we started another engine: Martin, p. 56
141 He remarked that: Insull, p. 333
141 Lights in any part of the house: *The Operator and Electrical World*, 21 April 1883
142 In all large cities: Private report by William Preece to Post Office
142 By this device: *The Electrical World*, 25 August 1883
142 It is so easy: *Boston Herald*, 10 January 1885

145 Thinking over the subject: Sir John Ambrose Fleming, *Memories of a Scientific Life*, p. 141
146 Everything: *The Electrical World*, 10 June 1899
146 By invitation of Mr. Edison: *Scientific American*, 5 June 1880
146 most pleasing: *New York Herald*, 25 June 1879
147 Mr. Edison is to the front: *The Operator*, 15 July 1880
147 will compare favourably: *The Electrical World*, 4 August 1883
148 I could not go on: *The Electrical World*, 9 August 1884
148 Mr. Edison today: *The Electrical Engineer*, 18 November 1891
148 Edison was honored: Roger Burlingame, *Engines of Democracy: Inventions and Society in Mature America*, p. 247 (afterwards referred to as "Burlingame")
151 women... do not seem: Mary C. Nerney, *Thomas A. Edison: A Modern Olympian*, p. 235
151 accomplished and serious: *Akron Times*, 4 March 1886
151 I asked her thus: Runes, p. 54
152 refreshingly independent: Dickson, p. 339
152 My father: Mrs. Madeleine Edison Sloane, in Souvenir Program of "Premiere Ball," held Orange, 1940, on launching of MGM film *Edison the Man*
153 See that valley?: Tate, p. 140
153 My ambition: Vanderbilt, p. 113
153 The method of developing: *Scientific American*, 17 September 1887
154 I lost the German patent: *Wall Street Journal*, quoted *The Literary Digest*, 13 September 1913
156 At one o'clock: *New York Times*, quoted *The Literary Digest*, 28 November 1913
157 Edison in his working gown: Young, p. 9
157 There is something quizzical: Young, p. 9
158 I know two great men: John O'Neill, *Prodigal Genius: The Life of Nikolas Tesla* (London), p. 60 (afterwards referred to as O'Neill")
158 There's fifty thousand dollars: O'Neill, p. 64
159 Tesla, you don't understand: O'Neill, p. 64
159 decided to compete: Harold C. Passer, *The Electrical Manufacturers, 1875–1900*, p. 167 (afterwards referred to as "Passer")
160 The electric lighting company: Edison, "Dangers of Electric Lighting," *North American Review*, November 1889
160 It is a matter: George Westinghouse—T. B. Ker, 11 July 1888, quoted Passer, p. 171
160 I'm very well aware: Edison—Henry Villard. Henry Villard Papers, Houghton Library, Harvard, quoted Passer, p. 174
161 could not understand: *Electrical Review*, 8 August 1903
161 I have been under: Edison—Henry Villard, 8 February 1890. Henry Villard Papers, Houghton Library, Harvard, quoted Passer, p. 104
162 But the visit: Tate, p. 136
162 Those fellows: Tate, p. 136
162 He then lowered: Gelatt, p. 38
163 At five minutes past two: *The Times*, 27 June 1888
164 The Devil's in it: *The Times*, 24 October 1931
164 Ah! Young lady: *The Times*, 27 October 1931
164 characteristic specimens: William Lynd, *Edison and the Perfected Phonograph*, p. 23
164 Ready?: *The Times*, 13 December 1890

164 If it were possible: Tate, p. 217
165 The partisans of Edison: Gelatt, p. 41
165 A doll which may speak: T. A. Edison, "The Phonograph and its Future," *North American Review*, June 1878
166 A large number of these girls: *Scientific American*, 26 April 1890
166 Hail! English shores: quoted Dickson, p. 135
166 be a dangerous thing: quoted Dickson, p. 220
167 entered into a contract: F. J. Garbit, *Edison's Speaking Phonograph*, p. 243.
167 A gigantic horn: Dickson, p. 137
167 When an extra-loud sound: Rosanoff
168 The public is very primitive: *The Etude*, October 1923
169 the transmission: *Scientific American*, 12 September 1914
170 In the year 1887: W. K. L. Dickson and Antonia Dickson, *History of the Kinetograph, Kinetoscope and Kinetophonograph*, p. 6
170 to weed out: Burlingame, p. 327
171 the idea occurred to me: *Century Illustrated Monthly Magazine*, June 1894
171 I think it will be admitted: *Nature*, 24 January 1878
172 Nothing was wanting: *Scientific American*, 5 June 1880
172 consulted with: Eadweard Muybridge, *Animals in Motion*, p. 4
172 Edison was enthusiastic: Kevin MacDonnell, *Eadweard Muybridge: The Man Who Invented the Moving Picture*, p. 31
173 I had a white cloth: Terry Ramsaye, *A Million and One Nights: A History of the Motion Picture* (London), vol. I, p. 58 afterwards referred to as "Ramsaye")
174 That's it: Ramsaye, Vol. I, p. 63
174 It made me: Tate, p. 246
174 filled to overflowing: Dickson, p. 251
175 I knew instantly: Albert Smith, *Two Reels and a Crank*, p. 80 (afterwards referred to as "Smith")
175 There was no screen: Smith, p. 80
175 If I made such an answer: Smith, p. 80
176 attracted quite a lot of attention: Runes, p. 77
176 No: if we make: Ramsaye, p. 119
177 If they exhibit: *New York Sun*, 22 May 1895
177 The lecturer dropped a china plate: I. F. Marcosson, "The Coming of the Talking Picture," *Munsey's Magazine*, March 1913 (afterwards referred to as "Marcosson")
178 The problem of actual synchronization: Marcosson
178 We used to experiment: Runes, p. 67
178 I had some glowing dreams: Runes, p. 63
178 The eye: Conversation between Thomas Alva Edison and John Philip Sousa, *The Etude*, October 1923
179 In order to carry out: Carl W. Ackerman, *George Eastman*, p. 221
179 I have never before seen: Tate, p. 260
179 Tate, if you want to know: Tate, p. 278
182 These few acres alone: Dyer and Martin, p. 481
182 In six or eight years: *Scientific American*, 2 April 1892
182 While I was experimenting: Dyer and Martin, p. 760
182 This was the day: *New York Times*, 26 September 1926

183 I never felt better: Dyer and Martin, p. 776
183 Then his face lighted up: Dyer and Martin, p. 505
185 The more extravagant: *Punch*, 30 October 1907
185 I have not gone into this: *Concrete and Constructional Engineering*, Vol. 2, 1907–8
185 in the style of Francis I: *Scientific American*, 16 November 1907
186 I noticed my old chief: quoted Woodbury, p. 213
187 I shall (then) turn my attention: *McClure's Magazine*, June 1893
187 by utilizing a vein: *The Times*, 25 September 1890
187 a magneto-mechanical device: *Scientific American*, 13 May 1893
188 I also found: *Century Illustrated Monthly Magazine*, May 1896
188 Edison would not consider: Dyer and Martin, p. 583
189 be able to inspect: *Electrical Engineer*, 3 June 1896
189 It is now said: *Pall Mall Gazette*, quoted Ray Allister, *Friese-Greene: Close-up of an Inventor*, p. 77
190 If Marconi says: Orrin E. Dunlap, Jr., *Marconi: The Man and His Wireless*, p. 248 (afterwards referred to as "Dunlap")
190 I would like to meet: Dunlap, p. 113
190 had thought that some time: Dunlap, p. 113
190 They are not needed: *Popular Electricity*, June 1910
191 He asked me no end of details: Ford with Crowther, p. 5.
192 he was now listening: Allan Nevins, *Ford: The Times, the Man and Company*, p. 167
192 As I began to explain: Ford with Crowther, p. 15
192 Beach: Dyer and Martin, p. 554
209 About 7:00 or 7:30: Dyer and Martin, p. 615
209 when we turned a sharp corner: *Electrical Review*, 8 August 1903
210 Sometime between midnight: Rosanoff
211 Ten revolving copper cylinders: Vanderbilt, p. 214
211 We're getting ready: *New York World*, 10 January 1914
211 I constructed a helicopter: *New York World*, 13 February 1923
211 You can make up your mind: Tate, p. 245
212 Within five years: *The Times*, 18 September 1908
212 flying machines: *New York Times*, 1 August 1909
212 If I were to build: *New York Times*, 1 August 1909
212 With a series of aeroplanes: *The Times*, 18 September 1908
213 Everything which decreases: *The Literary Digest*, 14 November 1914
213 five ounces to a meal: *The Literary Digest*, 14 November 1914
213 On Broadway: *The Forum*, December 1926
214 Where's your mother?: Charles Edison, Introduction to *Experiments You Can Do, Based on the Original Notebooks of Thomas Alva Edison*.
214 I knew: *New York Times*, 10 December 1914
215 Although I am over 67: *New York Times*, 10 December 1914
215 At the great Badesch: *Factory*, May 1912
216 the militarists: Allan Nevins and Frank Ernest Hill, *Ford: Expansion and Challenge, 1915–1933*, p. 312 (afterwards referred to as "Nevins and Hill")
216 However, a number of Edison's letters: Nevins and Hill, p. 312

216 something that is so practical: Warren
216 There are lots: A. L. Shands, *The Real Thomas Edison*, p. 12 (afterwards referred to as "Shands")
217 This was the only bed: *New York Times*, quoted *The Literary Digest*, 21 November 1931
218 beseiged by brokers: *Weekly Drug Markets*, quoted, *The Literary Digest*, 2 October 1915
218 I made application: *Scientific American*, 2 December 1916
218 Then there came an appeal: *Scientific American*, 2 December 1916
218 going into the war: *New York Times*, 22 May 1915
219 Even so logical: *New York Times*, 27 May 1915
219 Make her rock faster: *New York Times*, 10 May 1915
219 Three times a year: *New York Times*, 10 May 1915
220 make gunpowder and dynamite: *Scientific American*, 2 April 1892
220 It is a very pretty: *Scientific American*, 2 April 1892
221 Torpedo-boats: F. T. Miller, *Thomas Alva Edison* (London), p. 252 (afterwards referred to as "Miller")
221 I advocate: *New York Times*, 30 May 1915
222 I feel that our chances: Daniels, 7 July 1915, quoted Bryan, p. 229
222 replied with what the Navy: *New York Times*, 7 October 1915
222 You may rest assured: *New York Times*, 15 July 1915
223 The soldier of the future: *New York Times*, October 1915
223 Science is going to make war: *New York Time* 6 November 1915
224 They wanted: Luther Burbank, *The Harvest of the Years* (London), p. 223
224 They met the commander: Alfred Lief, *Harvey Firestone: Free Man of Enterprise*, p. 166
225 We made our first camp: *System*, May 1926
225 I did not know: *System*, May 1926
225 You must stay on board: Nevins and Hill, p. 39
226 Immediately upon meeting: *Science*, 15 January 1932
227 With this for a premise: *Scientific American*, 19 March 1921
227 This chart: Lloyd N. Scott, *Naval Consulting Board of the United States*, p. 167 (afterwards referred to as "Scott")
228 An electrician: Scott, p. 175
228 I made about forty-five inventions: *New York World*, 13 February 1923
229 simple, direct: *Science*, 15 January 1932
229 He was then: *Science*, 15 January 1932
231 With squads of newswriters: Charles E. Sorensen with Samuel T. Williamson, *My Forty Years with Ford* (London), p. 18
231 The day before the funeral: Miller, p. 255
231 When the doctor brings in: *New York Tribune*, 12 February 1921
232 So I made up my mind: *Scientific American*, November 1921
232 We were taken: Shands, p. 5
233 It may come some day: Runes, p. 91
233 I'm doing this: *Scientific American*, 19 March 1921
234 Are you one of the howlers: *New York Times*, 28 September 1921
234 That's the Opposition: Tate, p. 64

234 Can't go: Shands, p. 19
235 I don't claim: *Scientific American*, 30 October 1920
235 I cannot see: *The Forum*, November 1926
236 I think as you do: Woodbury, p. 336
236 It has to be in perfect order: *The American Magazine*, February 1930
237 They have spoiled everything: *The American Magazine*, February 1930
237 Don't make any mistake: *Popular Science Monthly*, December 1927
238 an annual crop: *Popular Science Monthly*, December 1927
238 Everything has turned to rubber: *The American Magazine*, February 1930
239 H'm, the same damn old: Nevins and Hill, p. 503
239 As (Edison) walked: *Detroit Free Press*, quoted *The Literary Digest*, 2 November 1929
240 Our floor: Nevins and Hill, p. 503
241 My message to you: Broadcast from Convention of the Electric Light Association, Atlantic City, 1 June 1931
241 Mr. Edison might be compared: *New York Time* 4 August 1931
241 It is very beautiful: Miller, p. 255
241 climbed the heights: Miller, p. 255

"New lamps for old."

Illustration
Acknowledgments

The following abbreviations are used:
ENHS Photograph supplied by the Edison National Historic Site, Orange, New Jersey
GEC Photograph supplied by the General Electric Research and Development Center, Schenectady, New York
IMP/GEH International Museum of Photography at George Eastman House, Rochester, New York
RTHPL Radio Times Hulton Picture Library, London

Frontispiece From *The Forum*, November 1926. Photo: British Library, London
6 From *Punch*, 25 June 1881. Photo: Derrick Witty
33 ENHS
34 Edison's birthplace, by kind permission of Madeleine Edison Sloane; Edison's parents, ENHS
35 ENHS
36 Edison seated at a telegraph key, ENHS; Grant's field telegraph, Western Union Corporation, New Jersey
37–39 ENHS
39 The Gold Room, Library of Congress, Washington
40–41 ENHS
42–43 Bell and his telephone, by courtesy of Bell Laboratories; The Manchester Telephone Exchange (from *The Telephone* by W. H. Preece and J. Maier, 1889) and Sir William Preece, by courtesy of the Post Office, London
44–48 ENHS
113 RTHPL
114 Sally Jordan's boarding house and the first photograph, GEC
115 ENHS
116–17 Edison's electric generator, from *Scientific American*, 18 October 1879. Photo: Derrick Witty; Edison Electric Light Co. and Edison Machine Works, GEC; Laying electricity cables, Library of Congress, Washington
118–19 Crystal Palace Exhibition, from *Illustrated London News*, 4 March 1882. Photo: London Electrotype Agency; "Jumbo" dynamo, GEC; Electric railroad, ENHS
120 Paris Exhibition, collection Georges Sirot; Sketch of light bulb, ENHS
121 ENHS
122–23 Glenmount, ENHS; Fourth of July, Mina Miller Edison, and fishing expedition, by kind permission of Madeleine Edison Sloane
124–25 Edison with phonograph and group photograph, ENHS; Gouraud dictating, his secretary, and recording session at Little Menlo, Christie's South Kensington; A distinguished gathering, from *The Times*, 19 October 1931. Photo: *The Times*
126 ENHS
127 Phonographs, Christies, South Kensington; Speaking doll, ENHS
128 ENHS
193 ENHS
194–95 Edison and Eastman, IMP/GEH; Edison with motion picture machine, ENHS; Film strips, William Gordon Davis; Kinetoscope, Science Museum, London
196–97 "Black Maria" and kinetoscope arcade, ENHS; Three films, IMP/GEH
198 Edison and Ford, RTHPL
198–99 ENHS
200–201 Edison and Naval Consulting Board, U.S. Navy photograph; Citizen's march and Flag Day, GEC; After the fire at West Orange, ENHS
202–203 Edison with his first grandchild, by kind permission of Madeleine Edison Sloane; Edison with his grandsons, and with his second wife, RTHPL
204–205 Camping trip 1916, ENHS; Camping trips 1918 and 1921, Firestone Tire and Rubber Company
206 GEC
207 Edison and Langmuir, GEC; Group at Dearborn, ENHS
208 ENHS
243 From the *New York World*-Telegram, 1931. Photo: The Institution of Electrical Engineers, London
250 From *Punch*, 24 December 1881. Photo: Derrick Witty

251

Index

Index

255